Endorsements for
Protecting the Places We Love

"Time is short to protect the 'last chance' ecosystems and crucial landscapes that provide habitat, protection from weather events, and connection with nature. Breece Robertson powerfully demonstrates how geospatial tools can guide critical decisions in the most urgent phase of conservation the world has ever faced. The Nature Conservancy is honored to have our partnerships and projects featured in this book that includes a wide array of maps, apps, and models for accelerating the pace and scale of conservation."

—**Jennifer Morris,** CEO of The Nature Conservancy

The Nature Conservancy (nature.org) is a global environmental nonprofit working to create a world where people and nature can thrive.

"The urgent global imperative to save our planet requires transformed policies, systems, and perspectives. It also requires democratization of data and analytic tools to enable all stakeholders and communities to participate in these efforts. This book could not be more timely. Breece Robertson provides both seasoned professionals and engaged citizens an essential guide for using GIS data and mapping that illuminates technical concepts and their practical application to address our most pressing issues related to land, water, climate, and social equity. An established leader and visionary in this rapidly evolving field, Breece is especially passionate about empowering diverse communities with advanced tools and technologies to achieve local solutions that contribute to global goals."

—**George W. McCarthy,** President and CEO, Lincoln Institute of Land Policy

The Lincoln Institute of Land Policy (lincolninst.edu) is an independent, nonpartisan organization whose mission is to help solve global economic, social, and environmental challenges to improve the quality of life through creative approaches to the use, taxation, and stewardship of land. The institute's Center for Geospatial Solutions provides people and organizations with tools to advance equitable solutions to social, economic, and environmental challenges.

CENTER FOR
GEOSPATIAL
SOLUTIONS

LINCOLN INSTITUTE
OF LAND POLICY

"Access to parks and green space is critical for improving health, protecting people from the climate crisis, and bringing communities together. Breece Robertson knows firsthand the transformational power of data when it's used to redress longstanding inequities in our communities. She also knows the urgency of acting today, as more and more people have too long been denied the healing benefits of nature. *Protecting the Places We Love* provides a brilliant road map forward for anyone who is curious about a world where fresh air, majestic landscapes, and human connection can help solve our most intractable social challenges."

—**Diane Regas,** President and CEO of The Trust for Public Land

The Trust for Public Land (tpl.org) creates parks and protects land for people, ensuring healthy, livable communities for generations to come.

PROTECTING THE PLACES WE LOVE

Conservation Strategies for Entrusted Lands and Parks

BREECE ROBERTSON

Esri Press
REDLANDS | CALIFORNIA

Esri Press, 380 New York Street, Redlands, California 92373-8100

Printed in the United States of America
25 24 23 22 2 3 4 5 6 7 8 9 10

ISBN: 9781589486164

Library of Congress Control Number: 2021932741

For purchasing and distribution options (both domestic and international), please visit esripress.esri.com.

Contents

Foreword

If you provide people with the right tools, they can explore and understand the world in new ways. With the right information, they can work more intelligently toward a positive outcome to make the world a better place. This is the philosophy and vision that has guided us since we started Esri® years ago. Our objective was, and continues to be, getting geographic information systems (GIS) into the hands of as many people as we can to give them the ability to transform data, any type of data, into remarkable, eye-opening information that will lead them to new paths of discovery and understanding. Along the way, we've met so many extraordinary people who share our vision and who, using GIS, have taught us so much about how the world works and what we can do to make the world a better place.

One of the extraordinary people I've met is Breece Robertson. At the time, she and a team of researchers and planners were working on a set of processes to advance land protection. I remember being especially impressed by her creative thinking. When she spoke at the Esri Geodesign Summit in 2018, she presented the Green Alleys project in south Los Angeles. The project mission was to transform derelict and underused alleyways into beautiful, usable pathways, where people could safely walk and play along paths filled with natural vegetation. Old alleys became climate-resilient corridors, a new kind of green infrastructure that could absorb storm water and bring pride to the community. It was inspiring.

Breece and Esri have worked together on many projects since 2002, including ParkScore, an analysis of the 100 largest city park systems in the United States. A partnership between Esri and The Trust for Public Land (TPL), ParkScore is a defining example of how to create park equity and need layers in GIS, combined with other geospatial and nongeospatial data, to create a first-of-its-kind ranking system for park and open space policy and funding decisions. In 2012, Esri awarded Breece the Making a Difference Award for her work.

In this book, Breece will guide you through innovative conservation strategies that you can employ in your organization. You will learn how to use GIS, strategic partnerships, and community engagement to improve the outcomes of conservation efforts, especially for parks and entrusted land projects. Breece will show you how to expand your use of GIS for strategic land protection and increase the pace and scale of the efforts needed before it's too late and some of our lands are lost forever. As Breece says, there is no time to waste, and we must act now to meet the scientists' goal of doubling conservation by 2030 to meet the enormous environmental challenges that we all face.

My husband, Jack, and I embrace bold strategies like those Breece presents. We hold these ideas close to our hearts and believe that people can make a difference if they explore and understand the unique and often hidden qualities of the landscape. Jack and I put our hearts into donating to preserve a magnificent 24,000-acre landscape at Point Conception, near Santa Barbara, California. In preservation, we find a sense of wonder and pride. It has opened our imagination not only to what can be, but to what should be for generations to come.

I encourage you to make your own difference, no matter how small or how grand.

—Laura Dangermond
April 2021

Acknowledgments

Writing a book that inspires and educates a diverse audience on how to use GIS for storytelling and strategy has been rewarding personally. It is inspiring to know about the efforts happening worldwide to save what needs to be saved. This book would not be possible without the experiences during my career, the support and partnerships from my colleagues in the conservation and park fields and beyond, and my friends and family. I am deeply grateful for the people who work hard every day to make our world a better place.

Thank you to Jack and Laura Dangermond for your passion and support for conservation and park efforts and for your friendship. Gratitude to the Esri team for partnering, brainstorming, and implementing conservation and park planning methodologies and to the Esri Press team.

Special thanks to Christina Kelso, Debbie Webb, Bill Webb, Robb MacLeod, Stephanie Smith, Ann Morgan, Jayne Miller, Heather Robertson, Bob Steimle, Diane Shanks, Sarah Noss, Holly Bostrom, David Weinstein, Julia Busiek, Lori Scott, Healy Hamilton, Regan Smyth, Sierra Arnold, Samantha Lee Belilty, Erin Heskett, Sylvia Bates, Anne Haven McDonnell, Yolanda Soler, Kent Ryan, Lynn McGlynn, Tyler Creech, Nick Viau, Jason Winner, Maureen Clark, Anne Scott, Jeff Allenby, Elyse Leyenberger, Will Rogers, Kathy Blaha, Brenda Schick, Ernest Cook, Ted Harrison, Chris Mathias, Jad Daley, Margie Kim Bermeo, Tree Webb, Lilian Pintea, Brandy Hurt, Valerie McNulty, Molly Robertson, Bill Leahy, Kelly Easterday, Mark Reynolds, Michael Scisco, Charles Rath, Deb Grieco, Mitchel Hannon, Sara Owen, Robert Kent, Will Allen, Christina Chelf, Kurt Menke, Lisa Johnson, Cheyenne Maddox, Laura Milton, Ryan Bahnfleth, Sanjiv Fernando, Lindsay Rosa, Joel Hanson, Linda Garrett, Connor Bailey, Liz Goldman, Joseph Kinyon, Melissa Clark, Kristeen Penrod, Susan Minnemeyer, Josh Richert, Michelle Cowardin, Larry Klimek, Levi Rose, Jon Osborne, Josh Ewing, Bryan Roeder, Liz Larter, Ben Jobson, Gak Stonn, Carrie Belding, Nicole Hance, Charleen Gavette, Bob Heuer, Brenda Faber, and Hilary Morris.

This list doesn't begin to include everyone whom I consider mentors, collaborators, supporters, and friends or include everyone who contributed in some way—you know who you are, and I thank you.

INTRODUCTION

Accelerating the scale and pace of land protection

The Eyes of the Future are looking back at us,
and they are praying for us to see beyond our own time.

—Terry Tempest Williams

Above, Milk Ranch Overlook Ruin, Cedar Mesa Area, Utah.

Josh Ewing, Friends of Cedar Mesa.

Every minute, we are losing our last wild places. Unprecedented human pressures replace verdant, precious, and unique places with pavement, buildings, and degradation. In the western United States, for instance, we are losing open space to human development at an alarming rate—one football field's worth every two and a half minutes and an area larger than Los Angeles each year (The Disappearing West 2019).

Once lost, most of these places can never be converted back to their natural state or integrated into a protected, connected system. It's not just the land we lose; we lose species, resilience during a time of climate crisis, and our connection to nature and to each other. These nature–human connections are critical to our individual and collective health and well-being, and to that of the planet. We lose habitat that supports the survival of species. We lose opportunities to provide communities with equitable access to parks and open space. We lose natural buffers and features that protect us from catastrophic climate events.

There is no time to waste: the earth is warming faster than scientists can measure and predict. People are losing their homes, communities, and livelihoods to processes exacerbated by climate change. We are losing species we haven't even discovered yet. Bold visions such as E. O. Wilson's Half-Earth Project and the Udall-Bennet Thirty by Thirty Resolution to Save Nature, by US Senators Tom Udall and Michael Bennet, have determined that we need to protect a minimal percentage of land to survive and thrive. Scientists agree that we must double conservation by 2030 to prevent catastrophic climate and ecological disasters that cannot be reversed. But we must act fast and scale our conservation efforts to meet the challenge. The year 2030 is fast approaching. Achieving our bold conservation goals will require strategic and targeted action—together. Only when we combine our efforts can we make the impossible possible. We each have a part to play.

The good news is that we have geospatial data, tools, applications, science, and the will of our communities to do smart conservation and park system planning. Today, this data, along with the methods and applications, is available to everyone, often for free online or through low-cost licenses or subscription services. These powerful tools have never been more accessible or simpler to use.

Data and community-driven approaches empower us with information that supports the urgent action needed to protect special places in danger of being lost forever. This information empowers us to create protected land systems that support species (figure Intro 1.1), habitats, natural resources, and healthy, livable communities that are climate resilient and socially cohesive.

For the purposes of this book, "land protection" and "protected lands" include all parts of a system of protected lands, from local parks to national parks, wilderness areas to urban parks, conservation easements, greenways and trails, and everything in between.

Whether you are a geographic information system (GIS) professional, CEO, executive, board member, land protection team member, or advocate, you can tap into the power of GIS to guide and implement strategic land protection for mission-critical impacts. This book is geared toward small- to medium-size land

Intro 1.1. Ackerson Meadows, Yosemite National Park—great gray owl.

trusts, conservation organizations, and park agencies that want to use GIS in expanded ways.

This book will provide field-tested approaches for creating or using maps, apps, and land protection planning methodologies. Drawing on my 20 years as a conservation professional, I'll explain different ways to approach challenges by using the power of storytelling and GIS analysis, including methods on how to map, model, and analyze specific land protection subjects such as biodiversity and park access. Pointers to tutorials, blogs, and case studies are included so you can dig deeper into any subject and learn how to do it yourself. The content ranges from creating maps for storytelling, analyzing park system access, equity, biodiversity, connectivity, climate resilience, large landscape conservation, and community engagement strategies to strategic conservation planning and more.

In each chapter, many of these questions will be explored:

- What is the challenge, need, or vision that GIS can help address or accomplish, and how do you approach it?
- Where do you get the data?
- What are the methods to map, model, and analyze issues?
- How do you translate the results into actions or recommendations?
- How are the results valuable, and why are the outcomes different because of GIS?

Included are tips and tricks, considerations, and examples from land protection organizations all over the globe. Many people think GIS is too expensive or too sophisticated and difficult to integrate, but that is no longer the case. GIS is becoming increasingly democratized through the move from desktop to web- and cloud-based computing. Any organization or department, big or small, can leverage GIS to support its mission and help guide its business direction. When you know what places need protection, you can create *bold* plans that leverage partnerships, inspire and engage communities, and produce lasting impacts.

On a personal level, I have always loved maps and nature. I spent a lot of time outdoors as a young girl, and I still do today. I built forts out of tree branches and pine boughs in the forest across the street from my house, fished in the pond down the road with my own fishing rod and tackle box, and brought snakes and salamanders home, to my mother's dismay. Today, I love exploring wild places, and I also love visiting nearby parks and open spaces where I can connect with nature, walk my dogs, and bike. I walk my dogs on the Santa Fe Rail Trail in Santa Fe, New Mexico, every morning and hike in the Galisteo Basin or the Sangre de Cristo Mountains weekly. I camp in remote, beautiful places on public lands monthly. I'm grateful that someone had the foresight to protect these places for me and for you.

Despite the challenges we face, I have hope for the future and determination to help solve some of our most urgent issues using GIS and technology. I'm passionate about making the world a better place by collaborating with others and bringing my skills, knowledge, wisdom, and dreams for a better future, in partnership with others, to make a difference. I chose a career in GIS because I was drawn to the opportunity to combine data, science, research, and community engagement with passion and heart to save and restore the places we love.

For 18 years, I had the great fortune of leading the GIS, research, and planning unit for a national conservation organization, The Trust for Public Land. My team developed deep partnerships with national organizations, federal agencies, local land trusts, community organizing groups, local governments, and others, and brought communities to the table to help lead GIS-driven land protection processes. I have firsthand experience developing conservation planning approaches, locally and nationally, and I have learned and refined my approaches along the way. I'm excited to capture this knowledge in a book to share with you.

This book is intended to educate and inspire organizations to use GIS to tackle the most pressing conservation and park issues. Global and local challenges are on the rise. It is urgent that we act now to bring all our tools and strategies to the table. I believe that we have immense untapped potential to accelerate land protection impacts globally. We can't afford not to use GIS to solve our biggest challenges. The time is now to embrace this powerful technology to scale conservation to save our planet. We have only one world. Time is of the essence.

References

Disappearing West. 2019. *The Disappearing West.* Accessed January 30, 2021. https://disappearingwest.org.

CHAPTER 1

How maps can save the places we love

Above, Galisteo Basin Preserve, New Mexico—rainbow.

Nevada Wier ©.

Maps inform. Maps tell stories. Maps ignite the imagination. Maps bring people together. And crucially, maps save places.

Storytelling with maps

Maps have helped us navigate and survive since the beginning of humankind. Telling the story of places with maps is an innate skill that we, as humans, have been using for a long time. Maps are etched into rocks all over the world by native peoples who lived in these places long ago and, in many instances, still do. Images carved into rocks are called petroglyphs, and some are communication tools, believed to show maps of landmarks and places to gather food, hunt, or fish. You can imagine how important these maps would be for navigation in arid environments where water is

Figure 1.1. Map Rock is a petroglyph believed to be the work of Shoshone-Bannock people before contact with American or European people and is thought to depict the surrounding Snake River region.

scarce or in dense forests or snow-covered areas where landmarks can shift and change. Historically, as people traveled the landscape, they needed information about the places they were moving through to help them survive.

Visual map images such as these petroglyphs helped people create cognitive maps of their surroundings that helped them navigate obstacles and have a deeper understanding of the landscape. The petroglyph in figure 1.1 shows a broad area around the Snake and Salmon Rivers in Idaho. It includes images of landmarks, animals, birds, and fish. Looking at it, I can imagine the indigenous people using the map to tell their own stories or to find healing herbs or the best areas to fish for salmon. Maps were important for storytelling and survival then, and they are just as important now for those same reasons—and more.

Today, our planet is in peril, and I believe that maps will help us navigate this precarious moment in time. We must identify and protect the places we love that are critical to our survival and that of the planet. When land is developed or degraded, we lose pieces of the fabric of our communities: our gathering places, plants and wildlife, species and ecosystems, climate resilience, the character of place, places for solace and solitude, places to grow our food, and the threads that connect us all. Protecting nature is protecting humanity and restoring balance to our world. But how do we help people understand what is at stake?

People respond to stories. If you want to make an impact, you must tell stories that are relatable and spark people to action. Today, as our technology has evolved from maps on rocks to maps on paper and maps delivered through the cloud, there are so many ways that maps can support and enhance our land protection stories.

Maps are magnets: unroll a map on the table during a meeting, and all eyes turn to it. With a quick scan, people can absorb and understand a place and an issue. Maps spark "aha" moments: "Wow, I never knew this place had so many threatened unique habitats for wildlife and birds!" Or questions: "If this property were developed, how would that affect water flow and water quality into our wild and scenic river and possibly affect my fishing or rafting business?" or "Why does my community lack parks and open spaces while this neighborhood over here has many?"

Maps put into context what needs to be protected and why. Everything is linked to place in some way, and maps get people on the same page about an issue, speaking the same language. Because maps allow us to capture and share a different kind of narrative, they help us tell the stories that are critical to protect and care for the places we love. Maps can support our struggles to protect the land by documenting history and change over time and providing continuity of information across generations. They play a critical role in the discourse of land protection, ongoing political litigation, and education.

The role of maps

- Maps provide context.
- Maps provide continuity.
- Maps document historical changes.
- Maps are visual storytelling devices.
- Maps provide a different kind of discourse for understanding issues.

Let's explore some examples of how maps have been used in land protection issues. The first example shows how maps helped save the Hoback Basin in Wyoming (figure 1.2). The second explores the quest of the indigenous tribes in the Four Corners region of the United States to protect the area, whereas the third describes how maps support investments in the park system in Pittsburgh, Pennsylvania. As you'll see, conservation and park organizations big and small can use maps as powerful tools to support their missions.

Figure 1.2. The Hoback River near the confluence with Granite Creek in Hoback Basin, Wyoming—Hoback Basin, Noble Basin, PXP lease area, 2012.

Conservation victory—the maps that protected the Hoback Basin

The Hoback Basin in western Wyoming is a key part of the Greater Yellowstone ecosystem, a landscape of beautiful mountains, vast forests, and abundant wildlife. It includes the headwaters of the federally designated Wild and Scenic Hoback River. Protecting this area is critical to water quality for western Wyoming communities.

In 2012, an oil and gas company decided to exercise drilling rights on their leases in the Hoback Basin and planned to create 30 miles of new roads and more than 130 oil and gas wells. The community was not included in the decision-making process. Locals knew this project would contaminate their drinking water, disrupt wildlife, create pollution, restrict access for outdoor activities, and destroy the area's wilderness solitude. Because the residents take great pride and enjoyment in the wilderness around them, the community knew they'd have to act fast to halt this project. So they reached out to local conservation partners to create a fund-raising campaign to meet the challenge. They needed to fully understand the project's potential impacts, defend their position using data and maps, inspire people to act, and raise the money to buy the leases.

Using maps, such as the one shown in figure 1.3, the campaign partners were able to help the community visualize where the drilling pads and 30 miles of new roads would be built. The maps showed the proximity of the project site to the Hoback River and the deforestation and fragmentation of the forest and habitats that would occur. Seeing these connections on the maps and imagining the potential destruction was shocking. The company owned drilling leases adjacent to the proposed project area and planned to expand into that area later. Other energy companies that owned leases in the vicinity were watching to see how the project unfolded—if it went forward, these companies would likely develop and drill soon. In so many ways, maps made clear just what was at stake.

Through the Save the Hoback campaign, the partners helped the community raise almost $6 million to purchase over 58,000 acres of oil and gas leases in the basin. The headwaters are now permanently protected. To get the full story, read *Saving Wyoming's Hoback: The Grassroots Movement that Stopped Natural Gas Development* (Shepard and Marsh 2017).

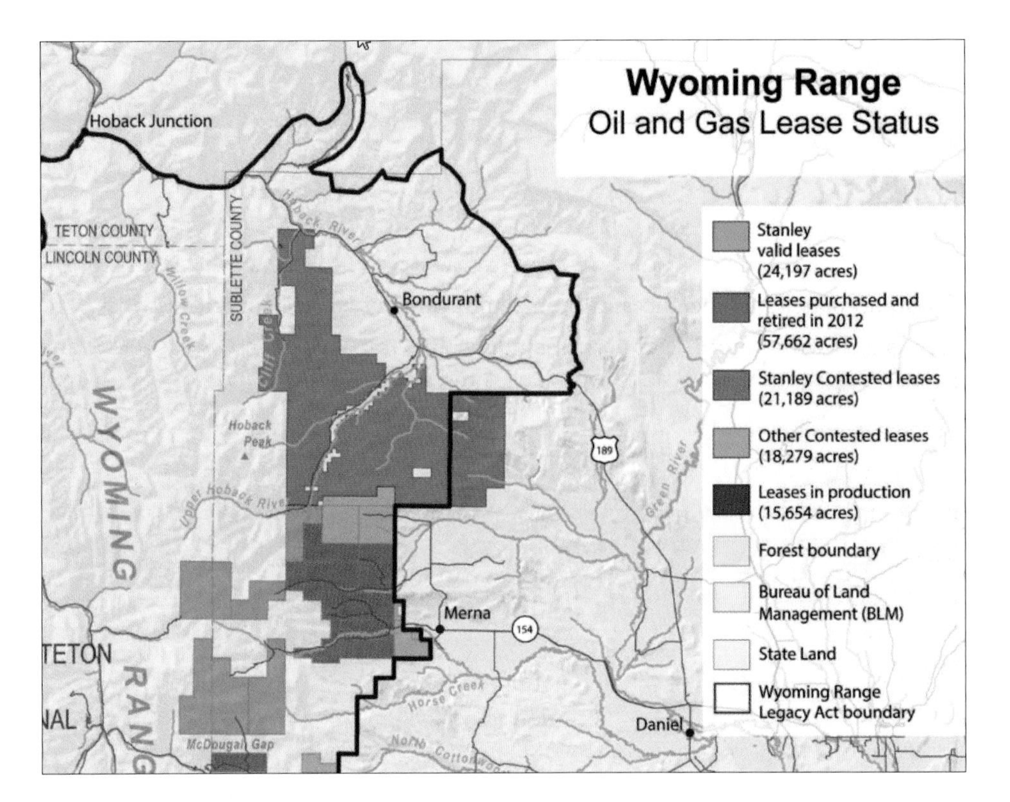

Figure 1.3. The Trust for Public Land acquired 24,197 acres (shown in bright green on the map) of oil and gas leases. The land, situated in the Noble Basin southeast of Jackson, Wyoming, is permanently protected from drilling, as are the blue and orange areas on the map. A full map of the Wyoming Range Legacy Act boundary can be found at JHNewsAndGuide.com.

What made maps so beneficial to protect the Hoback?

- Maps put in context the threats and all that was at stake then and into the future.
- Maps inspired the community to provide financial support to protect the Hoback.
- Maps conveyed the urgency with which the community needed to act.
- Maps told the story of the connection between people and nature in the landscape.

Creating indigenous ancestral homeland maps that reflect the people's relation to the land

The Four Corners area, located in the southwestern United States and part of the Colorado plateau, is the ancestral homeland to many Native American tribes. The area is rich with natural wonders, including the Grand Canyon, Mesa Verde, and Bears Ears. The area is thought to contain the highest density of archaeological sites in the nation. The cultural and natural abundance in this area has been at risk since the Western Expansion began in the 1850s. Native people have been cut off from many of their ancestral homelands and connections to the places that sustain them. Many of the cultural sites have been looted and destroyed, and roads and railroads have fragmented habitats and wilderness areas. Native people across the country are actively petitioning the government to address these wrongdoings that affect their lands, waters, health, and community.

In 2015, the tribes formed an intertribal coalition to petition the president of the United States to create the Bears Ears National Monument. The creation of the monument was designed to protect an area of 1.9 million acres that covered their ancestral lands (figure 1.4). Maps would become critical storytelling tools in this effort, but the maps had to reflect the views and relations to the land of the tribes in an authentic way. Maps can be both supportive and divisive—especially to peoples who have been displaced and divided from their lands by various mechanisms, almost all involving maps. This is where the boundaries that are typically placed on maps come into consideration.

Geographic information system (GIS) mapmakers create maps by typically layering contextual information such as local, state, and federal boundaries; protected lands such as national and state parks; and rivers and roads—all the "basemap" layers. But that approach did not accurately reflect the true intention of this effort to preserve Native American traditional knowledge. Bears Ears is a region where the knowledge and connection to the landscape are cultural, spiritual, and natural. These things cannot be fully captured by artificial lines drawn on a map that mostly have more political orientations than natural or cultural. The proposed Bears Ears National Monument was not a region developed around the boundaries created by politicians. The focus of the Bears Ears Inter-Tribal Coalition was based on a new paradigm of collaborative management between the tribes and federal agencies.

Figure 1.4. The *Region to the Native Eye* map depicts the proposed Bears Ears National Monument without state lines to graphically illustrate that the ancestral connection to Bears Ears transcends political boundaries. This map was created for the Bears Ears Inter-Tribal Coalition's proposal to President Barack Obama in 2015. Sometimes what is left off a map can send as powerful a message as what is included.

Stephanie Smith, Grand Canyon Trust.

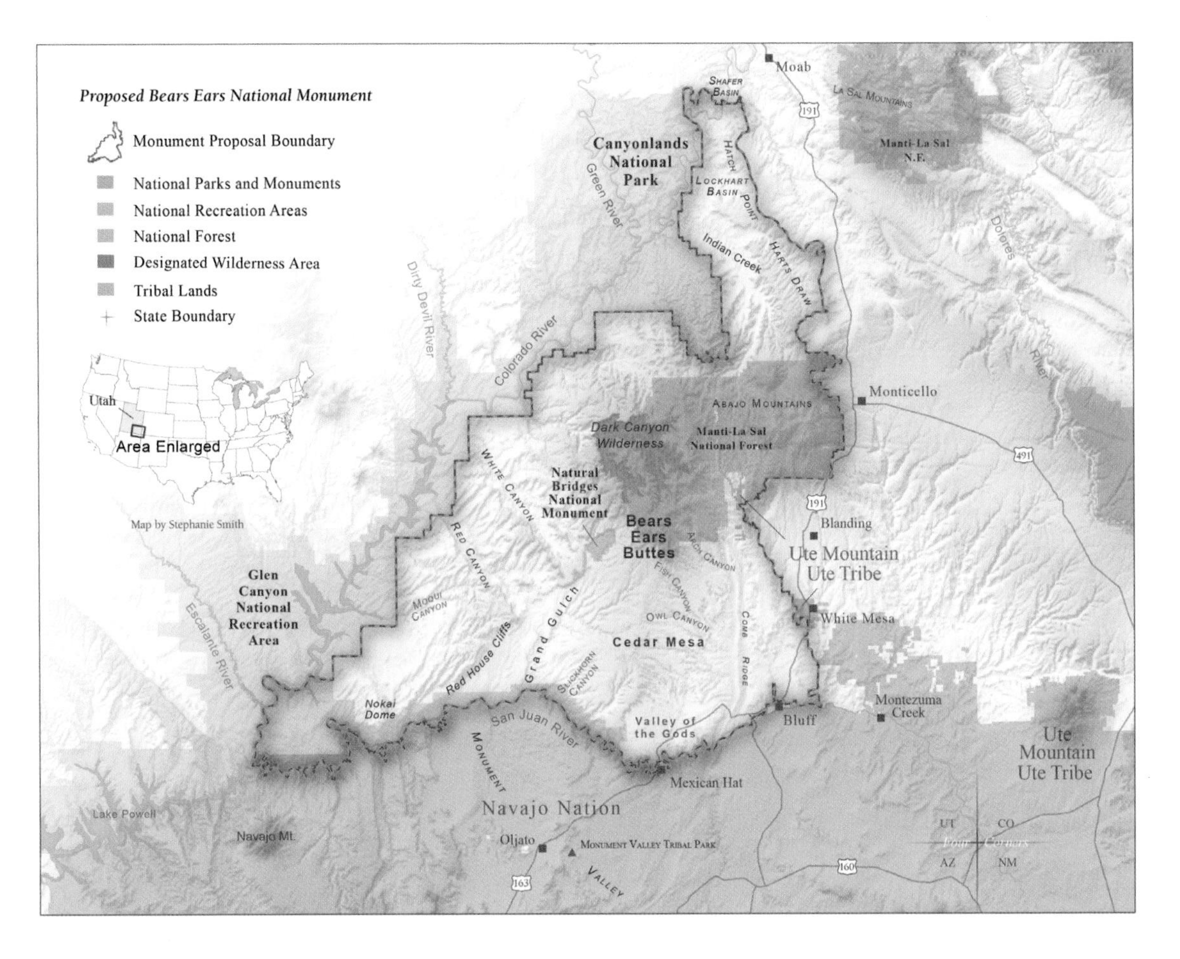

Figure 1.5. The *Proposed Bears Ears National Monument* map was used in the "Proposal to President Barack Obama for the Creation of Bears Ears National Monument" submitted on October 15, 2015. It demonstrates the physical landscapes proposed by the tribes for inclusion in Bears Ears National Monument. Instead of drawing the state lines as a hard boundary, the north arrow was used to illustrate the location of the Four Corners states. Although the arrow still shows where the state boundary lines are drawn, it downplays state line significance because the coalition does not see the region as bounded but as a landscape that defines their homelands.

Stephanie Smith, Grand Canyon Trust.

After creating initial basemaps drawn with those typical boundaries, the coalition decided to take a different approach in their monument proposal (figure 1.5).

The coalition led an extensive grassroots engagement effort, bringing together both elders with Native American traditional knowledge and younger people who continue traditional uses of the area, including hunting, holding ceremonies, enjoying solitude and beauty, and gathering herbs and medicines. Threats to the land and people were also identified. Thus, the resulting proposed boundary of the monument was not guided and driven by existing administrative boundaries but rather by the people and their own knowledge and longstanding relationship with the area.

Maps can be powerful, and maps can be hurtful. Emotions can be tied to maps. As storytellers using maps to support issues, we must recognize and consider the legacy and conflicts that maps have contributed to an issue or place. We cannot reverse the past, but moving forward, we can use maps informed by inclusive processes as integral tools to tell the story of the great things we can do together for a better future.

How maps and data guided an equitable park investment plan

Though people often think of parks in a one-dimensional way, simply as places to play and recreate, parks are a core component of what makes a community a healthy and enjoyable place to live. Parks provide many benefits to communities, such as improving physical and psychological health, providing places for people to socialize and gather, absorbing storm water, cooling down neighborhoods during extreme heat events, enhancing property values, providing habitat for wildlife, and much more. Yet many cities don't have the parks needed to fully serve their communities, and the parks that exist may need maintenance, repairs, or complete renovations to meet the needs of those wishing to use them. Parks must be redesigned and rebuilt on a regular basis. This has never been clearer or more relevant than during the coronavirus disease 2019 (COVID-19) pandemic, when parks were being used at rates exponentially higher than ever. Most cities struggle to find the money to create new parks and maintain the ones that exist, but there are good success stories for how cities can overcome this challenge.

The park system of Pittsburgh, Pennsylvania, includes 165 parks, parklets, and open spaces and ranks 15th in the nation, according to The Trust for Public Land's

ParkScore. Some 92 percent of residents live within a 10-minute walk to a park in Pittsburgh—but a 10-minute walk to what kind of park and in what condition? Disinvestment in the park system left many parks in disrepair, mostly in low-income neighborhoods and communities of color. The city has a $400 million backlog in capital projects that includes building new parks, renovating pools, repairing bike paths, and rebuilding athletic fields and playgrounds. It also has a $13 million annual maintenance funding shortfall to make the necessary repairs and provide upkeep for parks. The city wanted a new approach to address the needed park system investments and was committed to an equity approach, prioritizing the neighborhoods and parks most in need.

The Pittsburgh Parks Conservancy offered to help. The conservancy was created in 1996 by a group of concerned citizens who wanted to improve the park system. It recognized that the quality of the parks, especially historic parks, was declining. The conservancy formed a public interest partnership with the city to restore the parks. Over its 22-year history, the conservancy raised more than $112 million for parks, but more was needed to provide the necessary positive effects throughout the park system—especially in low-income neighborhoods and communities of color.

In 2018, the conservancy and the city made a commitment to bring all parks in Pittsburgh up to a high level of excellence. The conservancy and city worked together with the community to develop a long-term equitable investment strategy for the park system. This award-winning transparent and equitable park investment strategy combines the park need data for every park in the system with the community need data for every 10-minute park walkshed in the system and the public's priorities for investment.

The community engagement process, through a public survey, ensured that input and feedback provided by city residents represented the city's demographics geographically and in terms of income level, race and ethnicity, age, and households with and without children. The survey responses were collated, and the results were made available to the public citywide, by council district, by race, and by income level. This information drove the priorities for investment. The full complement of park and community need data, depicted in multiple venues, including an interactive map of the city, played a key role in telling the story, helping citizens understand the current conditions of the park system and the needs of every 10-minute walkshed of every park in the system.

The data depicted through the interactive map provided residents with the ability to graphically see neighborhood conditions across the city, such as obesity, depression, anxiety, diabetes, and asthma rates; poverty rates, including racially concentrated areas of poverty; and vacancy and crime rates across the city by neighborhood. The interactive map also allowed residents to see the scoring of every park in the system by park need and community need.

The questions that data and maps helped answer

- Which parks serve Pittsburgh's most vulnerable and historically underserved residents?
- Which parks sit in high-priority areas for improving tree cover and air and water quality?
- What is the condition of the parks?
- What types of investments are needed to raise all parks to the same level of quality?

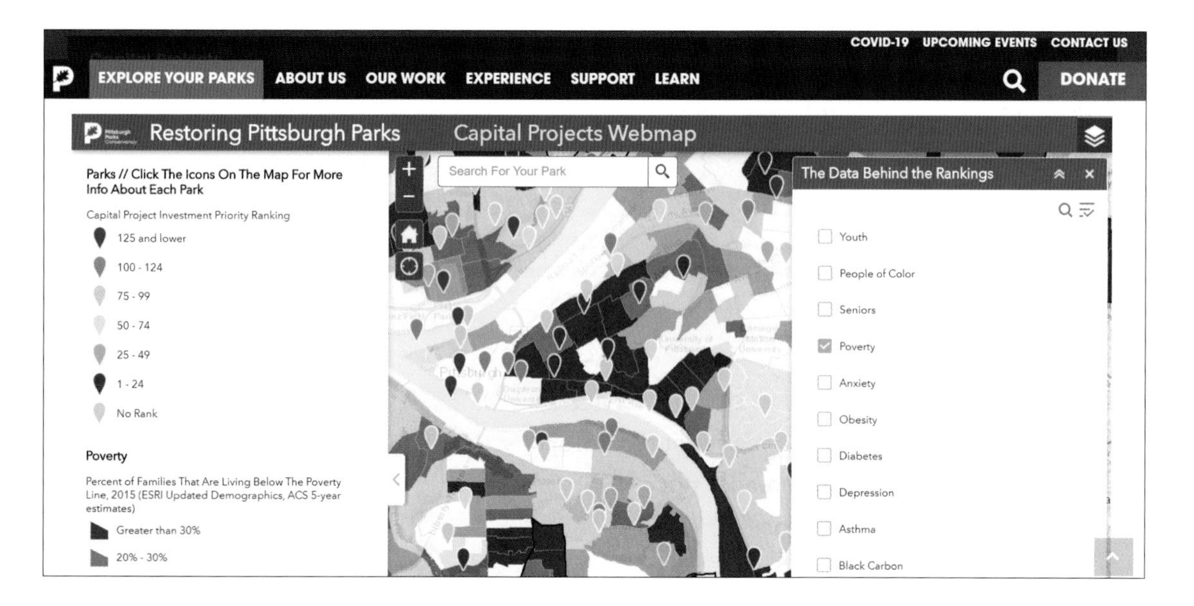

Figure 1.6. *Restoring Pittsburgh Parks Capital Projects* web map showing high rates of poverty in dark purple with parks that are ranked highest for needed capital improvement investments identified by the red markers.

Pittsburgh Parks Conservancy.

To prioritize sites for investment, the collaboration used an equity-based approach, analyzing more than 100 datasets related to four themes: people, community, health, and environment. This approach helped them identify the parks in the poorest condition with the greatest need for improvements. Overlaying data such as poverty, obesity, violent crime, tree canopy, and black carbon into an online, interactive map allowed community members to see how the parks in their neighborhood ranked for investments and improvements compared with others (figure 1.6). The publicly available maps made the data accessible to all and told the story about how an inclusive planning process combined with data would guide equitable park system improvements for Pittsburghers for generations to come.

How maps inspire and inform landowners

In my two decades as a conservation professional, I've joined many meetings with landowners who wanted to protect their land. Some were ranchers or farmers who were getting older and could no longer maintain their property. Some were passionate about preserving their land to protect wildlife and beauty. Some wanted to preserve their land for the broader public good, to protect water quality or cultural resources. But over and over again, when these landowners can't get the help they need to protect their lands, they are forced to sell some or all of their property. This is happening all over the world where families are forced to sell land for money, after which, usually, the lands are developed. These families care deeply about their land. Most could sketch it out on a piece of paper with as much precision and clarity as a cartographer. But many don't fully understand how their land fits into the bigger picture beyond the boundaries of their property. That's where maps can make a big impact in telling the broader story.

A map can show property boundaries, key habitats, water, and many other layers of information representing what the landowners already know. But additional maps that tell the story of the surrounding area can show how their property is a critical piece in a broader patchwork of surrounding lands—both public and private. During kitchen table meetings with landowners over maps, I've seen a deeper understanding and commitment to land protection begin to blossom. Many of these landowners get so inspired that they begin to educate their friends and fellow landowners about the importance of conserving their own lands. Maps can make their story come to life and transform their lived experiences into a geospatial intelligence about the importance of their property in the bigger schema of conservation in the area.

Maps that tell the story of organizational success

Most organizations understand the power of maps to support the stories of local land protection opportunities. But many organizations struggle with telling the story of the collective impact of all the places saved and parks created. Mapping completed projects, as shown in figure 1.7, can provide insights for land protection specialists to understand how conservation projects connect important places to one another or protect key puzzle pieces that are critical for habitat or providing public land access.

The GIS unit at The Trust for Public Land (TPL) spent nearly three years locating and mapping more than 5,000 projects nationwide at the tax parcel level. Once TPL had this information, stories about impacts and connections began to jump off the map—stories at all scales, from national to local. The patterns and relationships between TPL projects and other protected lands and initiatives were immediately apparent by looking at the maps. For example, clusters of lands protected by TPL within the context of lands protected by others showed how the collective efforts provided key linkages for wildlife movement or recreation opportunities for people. Adding data layers that show protected lands in context with natural resources, habitats, or regional initiatives can reveal a broader picture of what is needed or what benefits the current lands contribute.

In the southeastern United States, for example, overlaying black bear and longleaf pine habitat and range layers, along with the focal areas of the nonprofit, governmental, and for-profit entities working there, revealed that a coalition could be created to coordinate land protection efforts that addressed economic, environmental, and social benefits. Saving land for bears and pine trees, it turns out, is also good for economies and the health of the region. Maps are powerful visualization tools, but their underlying power comes from integrating the nature of geography, the connections across space, and the relationships and influences they can reveal.

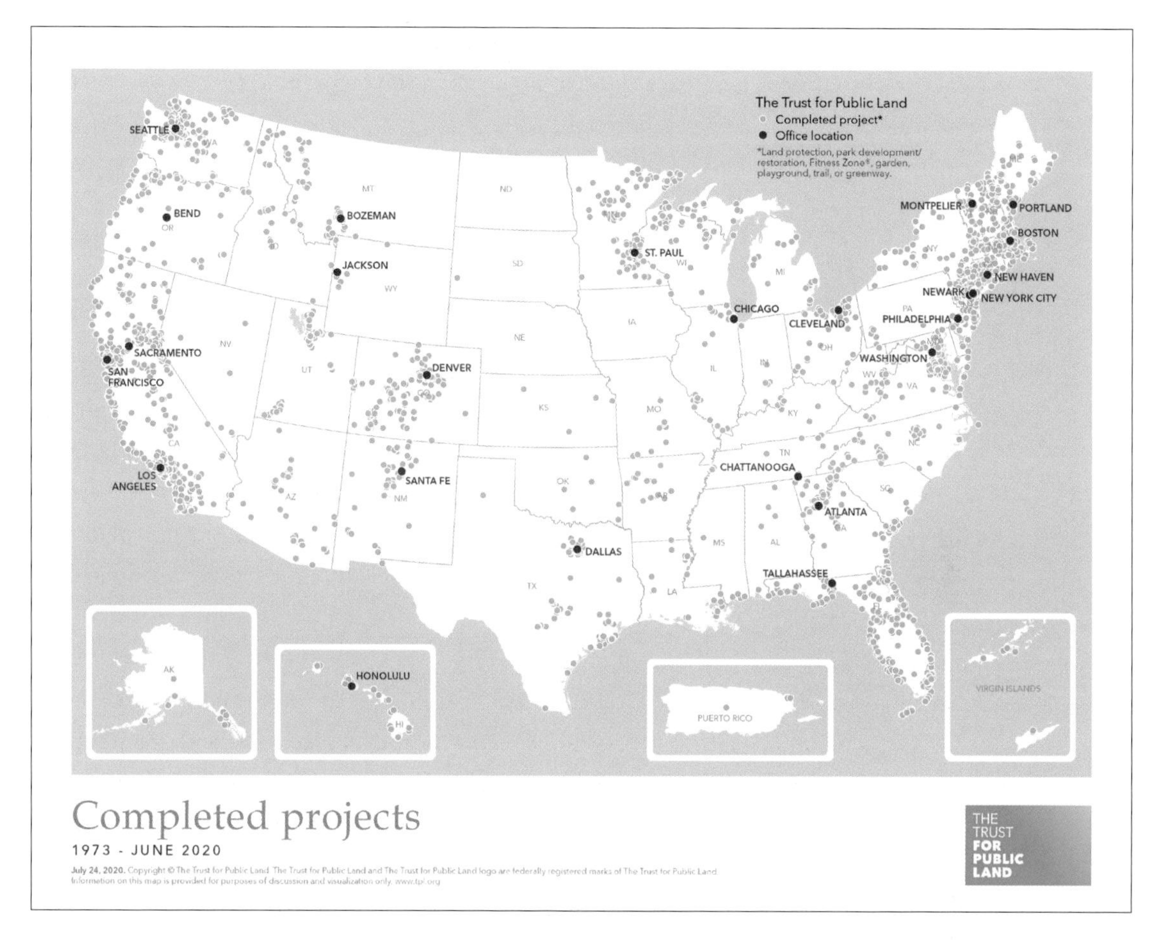

Figure 1.7. The Trust for Public Land's completed projects from 1973 to June 2020.

© The Trust for Public Land.

Maps that include state- or region-wide conservation projects in relation to spatial analysis layers give organizations a way to tell the story of conservation, which is a key step toward finding funding. Many nonprofit donors want to support work in their community or in their own backyard, and maps such as the ones in figures 1.8 and 1.9 can help them appreciate your conservation successes, understand where your organization is working, and identify focus areas for conservation activity that could develop into new programs or initiatives.

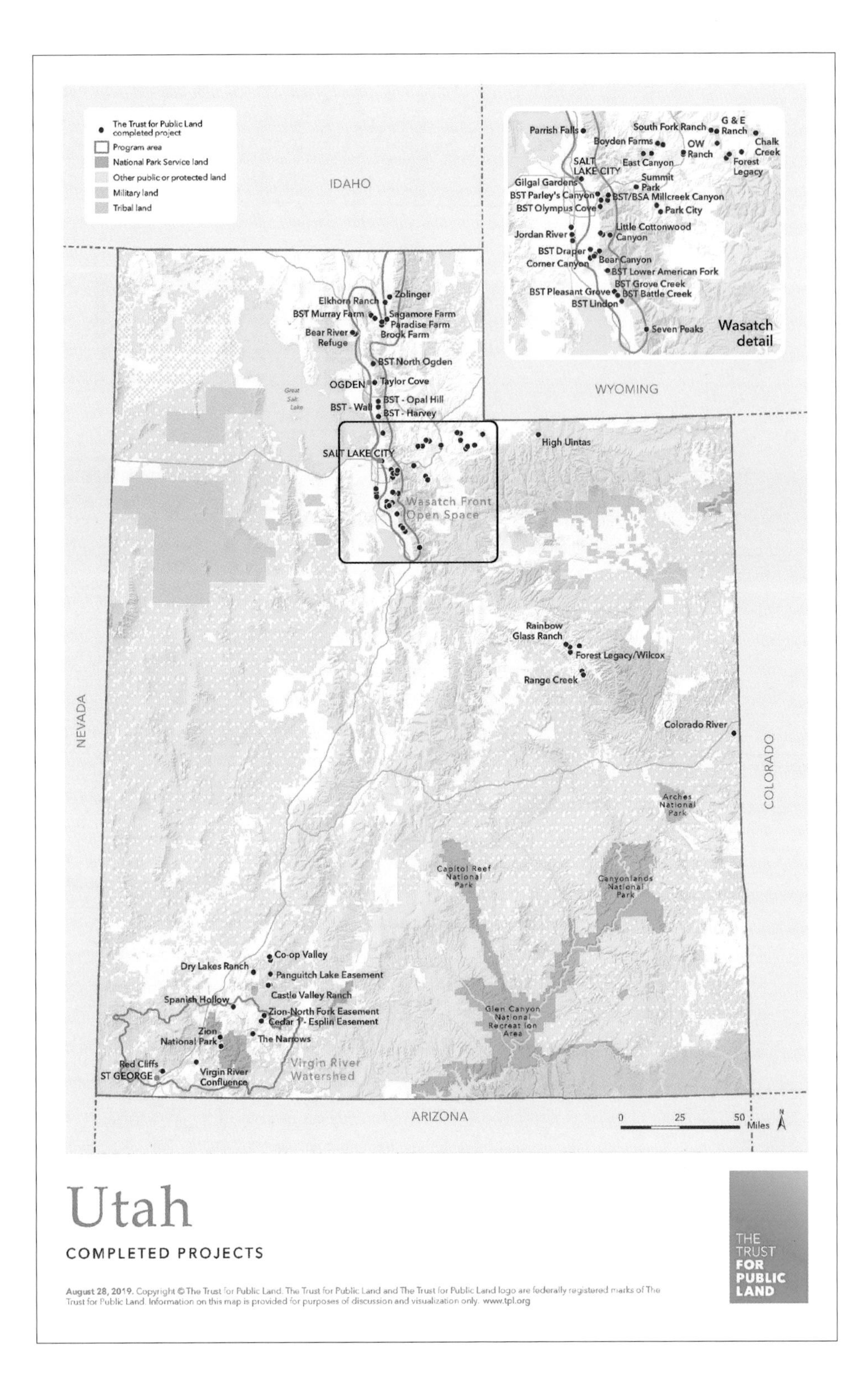
The Trust for Public Land completed project
Program area
National Park Service land
Other public or protected land
Military land
Tribal land
IDAHO
WYOMING
NEVADA
COLORADO
ARIZONA
Parrish Falls
South Fork Ranch
G & E Ranch
Boyden Farms
OW Ranch
Chalk Creek
Forest Legacy
SALT LAKE CITY
East Canyon
Gilgal Gardens
Summit Park
BST Parley's Canyon
BST/BSA Millcreek Canyon
BST Olympus Cove
Park City
Jordan River
Little Cottonwood Canyon
BST Draper
Corner Canyon
Bear Canyon
BST Lower American Fork
BST Grove Creek
BST Pleasant Grove
BST Battle Creek
BST Lindon
Seven Peaks
Wasatch detail
Elkhorn Ranch
Zolinger
BST Murray Farm
Sagamore Farm
Paradise Farm
Bear River Refuge
Brook Farm
BST North Ogden
OGDEN
Taylor Cove
BST - Wall
BST - Opal Hill
BST - Harvey
SALT LAKE CITY
High Uintas
Wasatch Front Open Space
Rainbow Glass Ranch
Forest Legacy/Wilcox
Range Creek
Colorado River
Arches National Park
Capitol Reef National Park
Canyonlands National Park
Glen Canyon National Recreation Area
Co-op Valley
Dry Lakes Ranch
Panguitch Lake Easement
Castle Valley Ranch
Spanish Hollow
Zion-North Fork Easement
Cedar 1 - Esplin Easement
Zion National Park
The Narrows
Red Cliffs
ST GEORGE
Virgin River Confluence
Virgin River Watershed
0 25 50 Miles
Utah
COMPLETED PROJECTS
August 28, 2019. Copyright © The Trust for Public Land. The Trust for Public Land and The Trust for Public Land logo are federally registered marks of The Trust for Public Land. Information on this map is provided for purposes of discussion and visualization only. www.tpl.org
THE TRUST FOR PUBLIC LAND

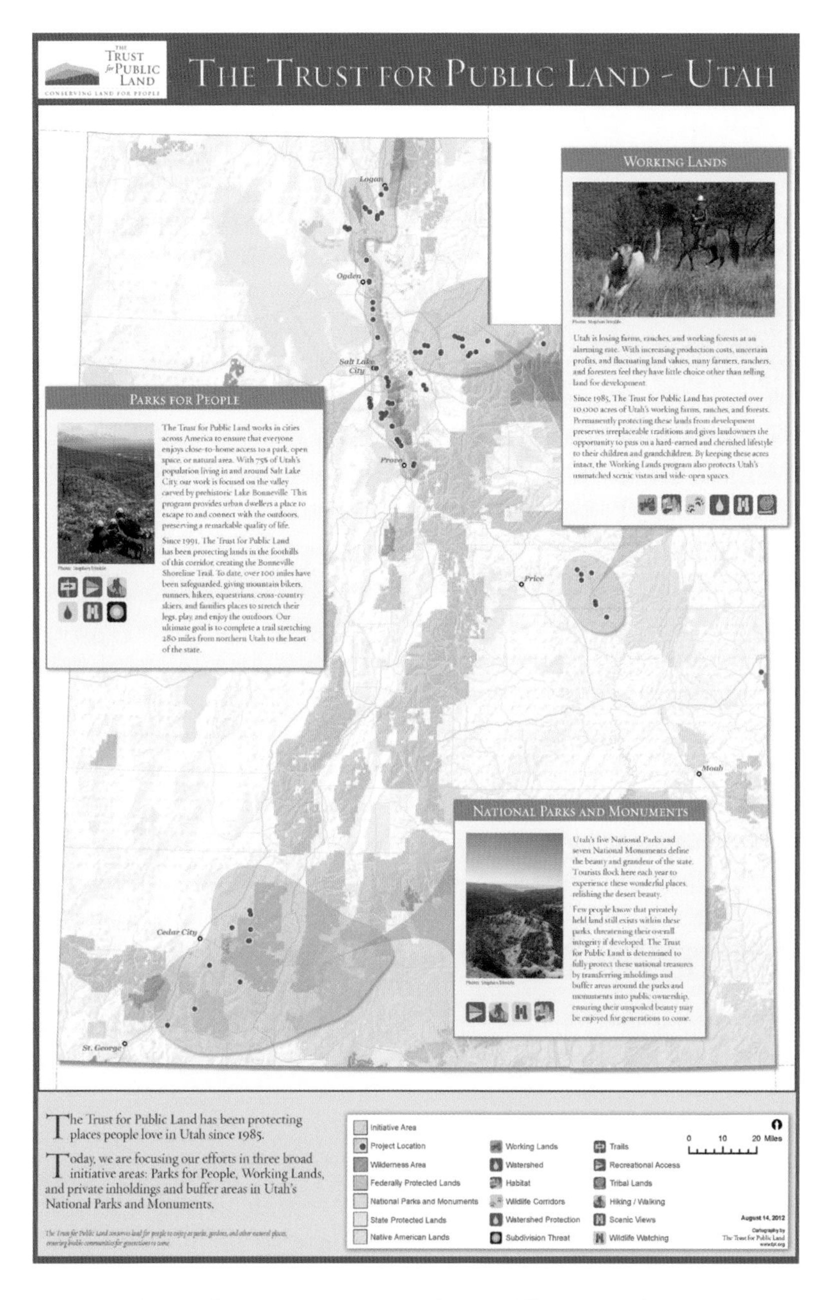

Figures 1.8 and 1.9. These maps enabled TPL offices to point to local successes and build out road maps for programs that focused on the conservation issues that needed the most attention locally. *Left:* This map shows completed TPL projects in Utah. *Right:* This map focuses on the story of TPL's initiatives within Utah.

In practical terms, how can your organization harness the power of the map to tell the story of conservation and park system benefits? How can maps uncover land-use patterns and organize efforts and energies around conservation priorities such as habitat protection or access? How can your organization use maps to understand how your work fits into both a local and broader movement to conserve our world?

StoryMaps stories

As you've seen, maps have evolved from being etched on rocks to being delivered through the cloud. ArcGIS® StoryMaps℠ is a cloud-based app that combines content such as interactive maps, photos, videos, and text into stories with engaging narratives and immersive experiences. For example, the viewer will scroll through or click on content to read the story, listen to a recording, watch a video, or pan, swipe, or zoom in an interactive map. Stitching together all these experiences provides a more powerful way to tell a story using a digital app rather than narrative, video, or maps alone. The story apps are easily configurable and can render complex information simply and powerfully. And because they are delivered through the cloud, this allows much broader distribution and visibility for the land protection stories we need to tell to engage, educate, and inspire people to act on behalf of park creation or conservation.

Author's note

In this book, I primarily refer readers to ArcGIS resources. There are several reasons for this, including that I have primarily used ArcGIS and partnered with Esri during my career to build data, analysis, and planning approaches and tools for sectors engaged in environmental protection. Another is that ArcGIS provides the most comprehensive resources for GIS. However, GIS in any form is valuable in the conservation field, and I encourage you to seek all GIS resources. In chapter 8, I expand the data and software resources to include freely available data and web apps and some subscription services.

StoryMaps stories can provide a tour of a conservation organization's protected landscapes, educate the viewer about the importance of a conservation goal, and more importantly, inspire them to make a commitment to conservation objectives. StoryMaps stories can be linked to your website, online conservation and park master plans, online newsletters, or news outlets. They are easily shared and searchable on the web and mobile devices and provide ways for community members to understand how they can get involved in conservation efforts. As conservationists, we strive to present technical results or data to the public in a way that is easy to understand.

As one example, the Defenders of Wildlife combines expertise in policy, law, communications, and GIS to bring transparency and understanding to complex issues involving large amounts of data and political jargon. The organization's Center for Conservation Innovation (CCI) uses tools such as StoryMaps stories and blogs to make geospatial information understandable and relevant to the public, as well as to journalists, activists, and lawmakers (figure 1.10).

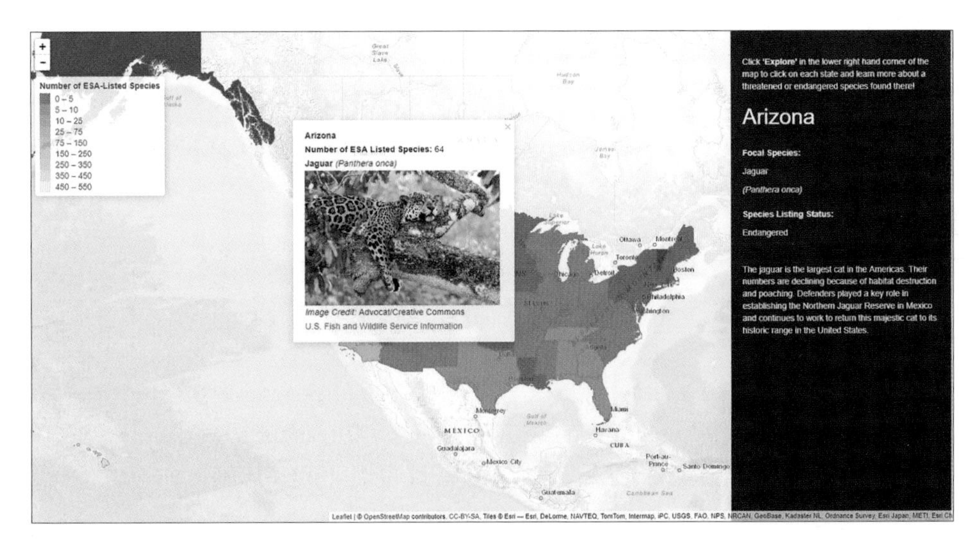

Figure 1.10. Screenshot of Defenders of Wildlife's StoryMaps story on changes to the Endangered Species Act regulations in 2019. The image is of an interactive map highlighting federally listed species in each of the 50 states. The example shows Arizona and the jaguar.

Figure 1.11. Defenders of Wildlife's StoryMaps story on changes to the ESA regulations in 2019. Image is of a Southern Resident orca, and text describes changes to the business-as-usual baseline of assessing impacts on listed species.

In 2019, the Trump administration changed Endangered Species Act (ESA) regulations, altering protections for threatened species and the ESA's science-based listing process. Defenders of Wildlife knew they had to act fast. They needed the public, policy makers, and the media to understand what was at stake. They gathered experts to decipher the new regulations and create an easy-to-understand narrative, and then tied it to place through a StoryMaps story (figure 1.11). The story showcased imperiled species at the global scale and in each state and territory that would be negatively affected by the new regulations. By focusing on how the regulations would impact species locally, the StoryMaps story made the information relatable and sparked action at all levels. The news media embedded the StoryMaps story directly into their articles, making it easy for the information to circulate rapidly and broadly.

Here's another example of an effective StoryMaps story, from Bernalillo County, New Mexico. The Bernalillo County (Albuquerque, New Mexico) Greenprint, a community-driven conservation and park plan, was completed in 2017. The purpose of the greenprint is to identify conservation and park opportunities that are supported and informed by the community and conservation partners. Data and GIS-based tools are used to identify where land protection will help protect unique resources such as cultural and historic sites, wildlife habitat, and local food

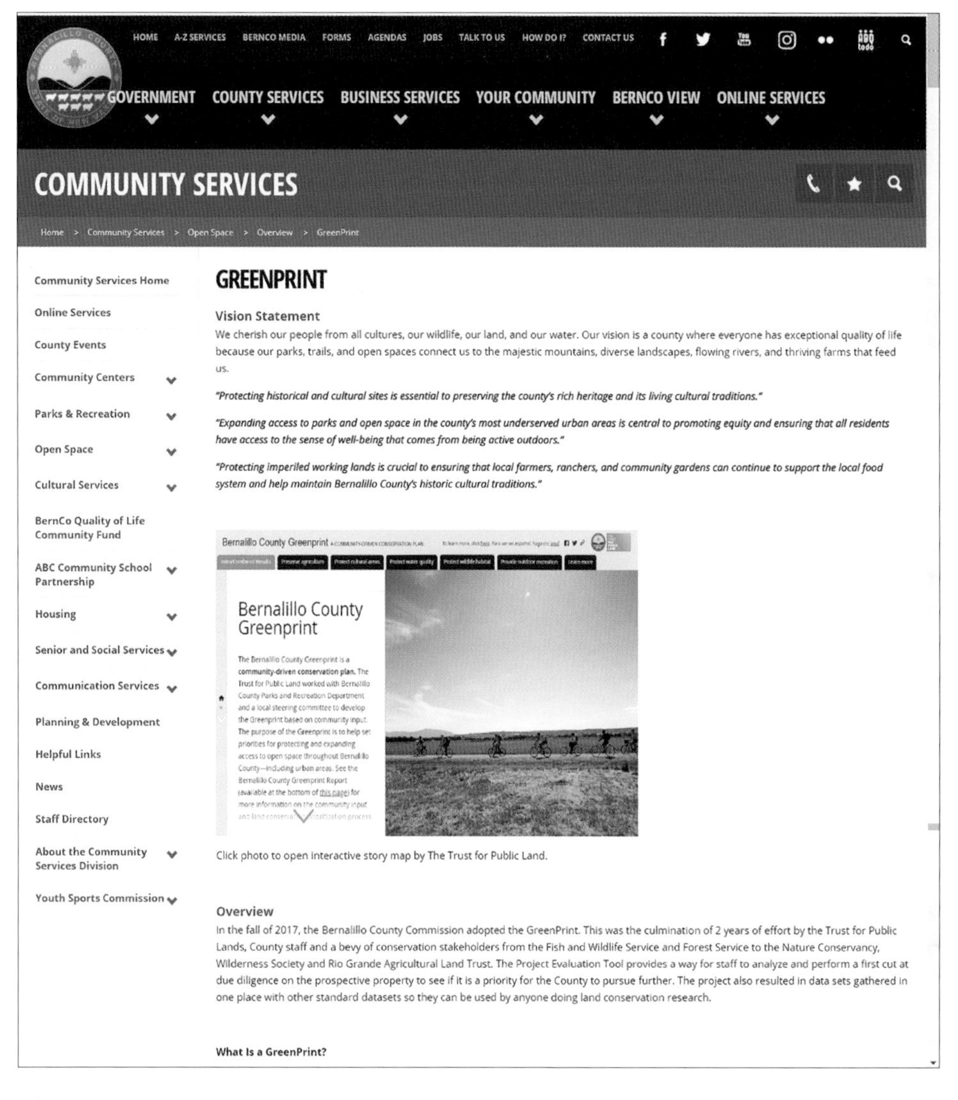

Figure 1.12. Bernalillo County, New Mexico, website showcasing the county's greenprint and the StoryMaps story.

Parks, Recreation and Open Space Department, Bernalillo County, New Mexico.

production, to name a few. The effort was led by the county and TPL and included many conservation stakeholders, including federal agencies and nonprofits. The online decision support tool compiles and displays data on many issues the community cares deeply about—such as conserving acequias, the historic ditch irrigation systems that cross New Mexico and irrigate fields while providing key trail linkages for wildlife, hiking, and biking. The decision support tool can be used by anyone but is geared toward planners and professionals. A StoryMaps story was needed to take the complexities of the information in the greenprint and translate it to be more accessible to the community (figure 1.12).

The StoryMaps story, available in English and Spanish, includes tabs in blue and black where the viewer can explore the conservation themes included in the Bernalillo County greenprint or conservation plan. The image on the right on the story's introductory tab (figure 1.13) shows an area near the Los Ranchos neighborhood in northwest Albuquerque. Red areas indicate natural lands along rivers, streams, arroyos, drains, and acequias that need to be protected to maintain continuity (figure 1.14). *Acequia* is a word that dates to when the territory was part of Mexico. Rich cultural traditions tied to acequias persist today, including communal water use, ditch cleaning, and sharing the responsibilities of water management. But many of the communal benefits can still be challenged. Imagine if a business purchased a property for a use that would pollute the acequias, and thus the agricultural fields they irrigate. Or a landowner might decide to cut down the trees along an acequia path that provide shade for trail users and nesting sites for migrating birds and wildlife. Identifying what needs to be protected and engaging the community to support the conservation plan is critical for success. StoryMaps stories make this data available and easy to understand for everyone from scientists to citizens.

Creating compelling maps that tell conservation stories

Knowing how to create maps and identifying where to get the data are the first steps to using GIS for conservation. This section covers both.

Creating a great map is not a simple task. Understanding how to organize and style data to tell a compelling story is a skill that requires training in GIS, a good eye for design, and a focus on who will be reading the map and what you want them to take away from it. Stylized map types range from printed maps to interactive apps that help you visualize your data, tell your story, and perceive things that you couldn't do otherwise. Stylized maps are best for sharing broadly, to include in marketing or information packets or on websites. These maps should incorporate the best cartographic design principles, graphics, and techniques for portraying the best stories. (See "Resources for Mapping" later in this chapter for more on these

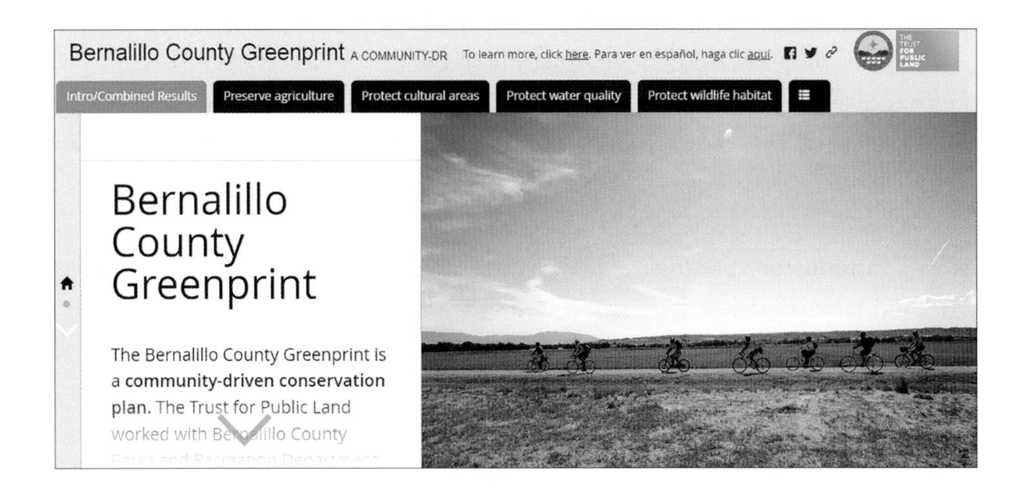

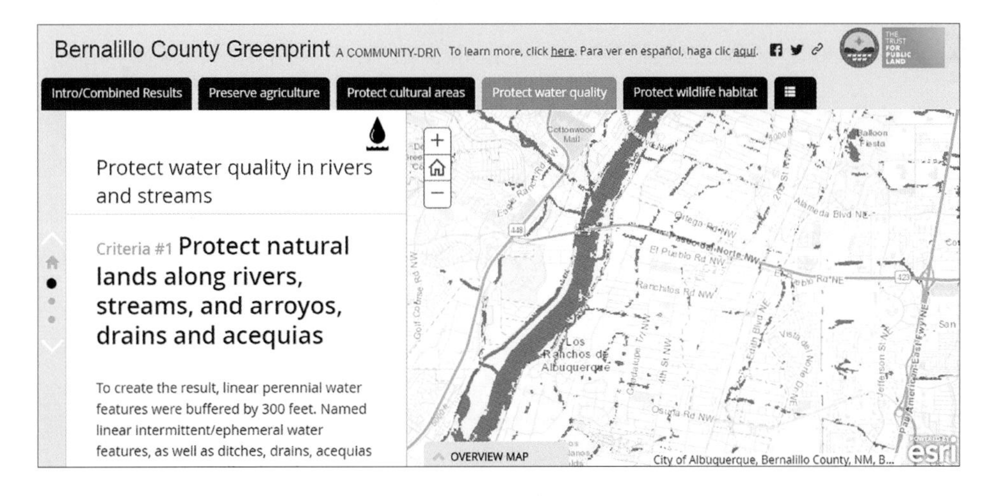

Figures 1.13 and 1.14. Bernalillo County (New Mexico) Greenprint StoryMaps story developed by TPL. *Top:* The story's introductory tab. *Bottom:* The water quality tab and areas in red on the map that should be preserved for that purpose.

principles.) You can work with partners, universities, GISCorps volunteers, or consulting firms. You can also take advantage of simplified online GIS tools to create interactive maps to tell the story of your work.

Next, we'll cover some areas to consider before investing in in-house GIS capacity.

How many maps do you need?

Suggested maps for external use

- A map of the geographic focus of your organization
- A map that shows what your organization has accomplished, such as land protected, programs or initiatives, species reintroduced, geographic focal areas, money raised, or however you define impacts
- Individual maps for impact areas such as climate resilience or agricultural preservation
- Current land protection projects if suitable for external communication. If not, these can be kept internally until the time is right to share publicly but still be used to inform, educate, and inspire action.
- Program or initiative maps

Suggested maps for internal use

- Potential projects being considered but not yet suitable for public display
- Map of your donor base—current and potential
- Program or initiative maps
- Maps of partnerships or collaborations
- Maps that show the characteristics of your focal area, such as demographic and natural resource information that can help guide strategic decision-making

Important tips and considerations

- Branding your maps—organizational maps explain what your organization does. Make sure your maps are stylized and branded to distinguish your organization's unique position and role.
- Internal or working maps versus final map products—keep in mind that what you publish reflects your organization's brand and identity. Make sure that maps created for internal use don't get picked up and used for unintended purposes. (I've watched maps intended for internal use end up in the Sunday newspaper.)

- Map costs—creating a stylized map for external use will likely take three to four times longer than a simple map for internal use. It takes anywhere from under a half hour to create a quick contextual map to eight-plus hours for a branded, stylized map prepared for external use. These maps cost anywhere from under $100 to thousands of dollars to create.
- Data costs—most data needed to create maps will be free, but some, such as digitizing (drawing) your new project boundaries or purchasing an updated parcel data layer, will require time and resources.
- Map updates—some maps need to be created only once without updating, and others will need to be updated frequently. You can update your completed projects map quarterly, biannually, or annually depending on the number of projects completed. Align map updates with your organization's calendar for board meetings, fiscal year-end reporting requirements, foundation and grant reports, and website updates or changes.
- Owning your data and map files—when you engage a consultant or partner to create data or maps, create a written agreement stating that the deliverables, including digital copies of the input data, GIS project files, and map image files, belong to you. Even if you don't have a GIS specialist on staff, you may need those digital files in the future, whether you decide to hire a GIS specialist or work with a different consultant who will use those files for updates or new maps.
- Digital or hard-copy maps—some of the maps you create for hard-copy distribution can easily be turned into interactive maps to display through your organization's website via a simple API. You don't need to pay someone to write a custom mapping app for your organization's website. Esri® APIs allow for seamless access to interactive maps of your focal area, your projects, and information about projects. The data and maps can be stored on a remote server and referenced via links or directly embedded into your organization's website. Esri continues each year to make

the creation and use of apps more accessible. Why not have both hard-copy and digital interactive maps and apps that tell the story of your organization and vision?

Resources for mapping

Following are some resources that can inspire and develop your map-making skills.

- GreenInfo Network is a California-based nonprofit organization that provides GIS services for nonprofits and agencies. The company has a website, Land Trust GIS, which provides best practices, examples of land trusts using GIS in various ways, and supportive resources, including essential maps for every land trust.
- *Cartography.* by Ken Field (Esri Press, 2018).
- *Map Use: Reading, Analysis, Interpretation*, eighth edition, by A. Jon Kimerling, Aileen R. Buckley, Phillip C. Muehrcke, and Juliana O. Muehrcke (Esri Press, 2016).
- *The ArcGIS Book* at TheArcGISBook.com includes a lot of information on how to get started with mapping and analysis. It includes clickable and interactive maps.
- The ArcGIS StoryMaps Gallery on esri.com showcases examples of user stories and StoryMaps templates that you can access for inspiration and guidance.

Finding and using data

Data is more abundant and available than ever before. (Most chapters in this book have information on where to find data and how peers have used data for their efforts. Chapter 2 includes a section on finding protected areas data.) GIS is more accessible to a wider audience through ArcGIS® Online, open-source, and other cloud-based apps. Data science and computing power have evolved rapidly, making big data accessible for precision conservation. We can even access anonymized cell phone data today for a fraction of the price it would have cost a year ago. This data can be used to understand how many people are visiting public lands or parks, where they are traveling from, and other information that can help with outreach and inspire support.

When making a map, your first consideration is where to find data, how to prepare it, and how to use it. For example, will your map need data on water sources, wildlife habitat, and farmland? Data is stored on servers and computers in many places, sometimes called "data silos." Most cities and counties have GIS departments, and some share data freely through their websites, whereas others provide data for a fee. Some cities and counties require you to contact them directly and request the data, which can be delivered online or, in some cases, on an external hard drive. Increasingly, municipalities and others are serving their data through open data portals that make it easy to freely download the data you need.

Data can be stored in cumbersome formats that are hard to access and understand. Spreadsheets, large reports, complex databases, and other forms of tabular or narrative data require interpretation and tools to understand the information and the stories and insights the data reveals. Software such as Tableau, a Salesforce company, and ArcGIS® Insights℠ help you access and interpret nongeospatial data. All you need to link nongeospatial data to a GIS map is a location field in the attribute table. Examples of location fields include zip codes, coordinates (latitude/longitude), addresses, and parcel numbers. Database experts can help you access data

Tip

Esri has a variety of programs offering discounted GIS software licenses for personal use and nonprofit organizations at https://www.esri.com/en-us/industries/sustainability/nonprofit-program/overview. Another option is through the Society for Conservation GIS (SCGIS), an all-volunteer community of GIS users. The organization assists people and organizations worldwide working in the field of conservation through communications and networking, including an annual conference. It also provides scholarships and training. All active SCGIS members can get access to an organizational ArcGIS Online account (Creator level). This account offers much more functionality than the free account. Being a member of SCGIS provides many benefits, including support through a talented and passionate community of conservation professionals.

from many locations that require data connections, export/import, and relational processes.

If you are just getting started, go to ArcGIS Online and explore the multitude of data available from many public and private sources. ArcGIS Online is a cloud-based solution that supports creating maps, doing analysis, sharing maps with others, and collaborating. You can access ArcGIS Online through web browsers or mobile devices and through ArcGIS apps. You can create a free, public account that allows you to create maps and share them publicly. Organizational accounts provide more functionality, including more ways to share content and data. To learn

Tips and tricks: Parcel data

Land parcel data (or cadastral data) includes information about properties and associated attributes such as taxes or zoning and is important in land protection mapping. Digital parcel data doesn't exist for every place, but availability is becoming more widespread. In the United States, parcel data exist for almost the entire country, but it is managed and sold by companies and can be expensive. If you need to create a map that includes properties in your city or county, contact your municipality or state and ask for information on how you can access parcel data through the assessor, planning, or GIS department. If the parcel or property data don't exist for your community, you or a consultant may need to digitize or draw that data. Once you have parcel information, you can easily calculate information such as parcel acreage and perform other analyses, such as calculating how much of a certain habitat is within the parcel boundary or within a distance of that parcel, or how many people visit the property for recreation within a certain distance. Once you have the property boundary in a geospatial format, you can begin to perform geospatial analyses to tell the story of why it's important to conserve.

If you don't need highly stylized printed maps but you need to visualize your own data or access content from Esri or other organizations, use an online mapping app such as ArcGIS Online, which allows you to upload and stylize your own data, integrate data from other sources, and share your maps with others. Ask to join an organization's ArcGIS Online group or create your own personal account.

how to set up an ArcGIS organizational account, go to learn.arcgis.com and search for the lesson on setting up an ArcGIS organizational account.

The ArcGIS Online website, arcgis.com, offers guidance on how to get started using ArcGIS Online and provides many other resources. Hundreds of thousands of public datasets are managed and shared in ArcGIS Online, including data from many government agencies. This data is generally formatted and ready to be used in your map with the click of a button. How do you know which datasets are reliable and authoritative? Esri is doing a lot of work in authenticating and curating data. More than 8,000 datasets available through ArcGIS® Living Atlas of the World are curated by Esri and deemed reliable and authoritative. We'll explore many conservation data sources in this book. Chapter 8 is dedicated to online data and mapping apps, most of which are freely available.

Resources on how to make maps

Learn ArcGIS® provides guided lessons, tutorials, and resources for all types of GIS uses and applications. For step-by-step instructions on how to create a web map and an interactive web app, start with the "Get Started with ArcGIS Online" on learn.arcgis.com. Explore the other topics on this page, and try some of the lessons that focus on creating and exploring maps.

More resources for working with data to create maps

- *The GIS Guide to Public Domain Data* by Joseph Kerski and Jill Clark (Esri Press, 2012).
- The Spatial Reserves Blog on WordPress has an array of resources for getting started with and understanding geospatial data using hands-on exercises.
- Universities with GIS programs have GIS or cartography resources on their websites. Membership groups such as the British Cartographic Society offer a wealth of information on cartography.
- Websites of your favorite conservation groups often have beautiful maps that you can use for inspiration in developing your own.

If you are an experienced cartographer or GIS professional, try the lessons in Learn ArcGIS for ArcGIS® Pro, ArcMap™, and ArcGIS® Enterprise℠ by selecting the mapping capability of interest. Here you can also create a Learn ArcGIS lesson that will support and help others in their quest for conservation insights and action through maps.

Resources for creating StoryMaps stories

To learn all about StoryMaps, search for and explore the ArcGIS StoryMaps web pages on esri.com. Here you will get information on how to get started. You can view the StoryMaps gallery to get inspired and access step-by-step approaches on how to make your first StoryMaps story. Creating StoryMaps stories requires planning, creating the narrative, creating the map or maps that will support the story, iterating through the design, and then sharing the story. The resources on esri.com will walk you through these steps.

Learn ArcGIS includes many articles and lessons on how to get started with creating StoryMaps stories for the beginner and how to enhance and expand the creation and use of StoryMaps.

Accessing GIS training resources

Whether you're looking to gain more skills individually or train others in your organization, Esri provides many education options to fit different levels of knowledge. It also has instructor-led trainings, e-learning, tutorials, and videos. These training resources can help you learn how to use GIS for your organizational needs, whether simple or complex. Search for "training" on esri.com. SCGIS is also an excellent resource for training (see the tip on page 33).

Challenge

If you are a GIS analyst or enthusiast, learn about the various workflows and issues your colleagues experience and brainstorm with them on how GIS-based solutions could help. Have them join you at your computer and show them different data layers, how they overlay, and how you can join an Excel table of conservation projects to a geospatial layer to visualize results on a map immediately. Whether your organization is big or small, GIS offers opportunities to improve efficiency and effectiveness. Keep track of the examples in which GIS helped to report to your manager and to the board. You may be surprised by how many problems GIS can help your organization solve.

References

Shepard, Florence R., and Susan L. Marsh. 2017. *Saving Wyoming's Hoback: The Grassroots Movement that Stopped Natural Gas Development*. Salt Lake City: University of Utah Press.

CHAPTER 2

Enhancing urban park and green space systems: Access, equity, biodiversity, and connectivity

Above, Lakeview Terrace, Ohio.

What if parks could help our communities withstand climate change by replacing pavement with grass, trees, and modern green infrastructure elements that absorb storm water and keep the neighborhood cooler and more comfortable on hot summer days? What if every community had a well-connected park and green space system that provided benefits such as habitat and connectivity for species and recreation opportunities for people? What if every student had access to a fun, exciting green playground instead of an asphalt or dirt lot (figures 2.1 and 2.2)? What if every person, in every city and town, had a park within a short

Figures 2.1 and 2.2. The playground at PS 33 Edward M. Funk Elementary School in Queens Village, New York, *top*, was once just a blank concrete lot, subject to flooding during storms and scorching heat in summer. In 2010, the school community got together to redesign their outdoor space and dreamed up a welcoming, healthy playground, *bottom*, that's open to all.

walk of home? What if those parks were designed and built to enhance quality of life for everyone in the community?

This vision of a greener, more equitable future for every community is within reach. Achieving it will require cooperation among experts, elected leaders, and concerned citizens, along with strategic public investments—all of which can be informed and driven by data and the community conservation processes we are using today. By using these processes, we can guide planners and park professionals to site and construct parks that work better for everyone.

Benefits and challenges

Imagine life without access to your favorite park or open space: no place to get fresh air, exercise, reconnect with nature, or commune with your friends. No place to sit in the cool shade of a tree on a hot summer day. For too many people, this is the daily reality. In the United States, more than 100 million people—one in three Americans—do not have a park within a 10-minute walk of home (The Trust for Public Land 2015).

In some cities, it's even worse. In Los Angeles, two out of three kids don't have access to a park (figure 2.3). Research shows that close-to-home parks improve health and strengthen communities. A study by the UCLA Center for Health Policy Research (Babey et al. 2013) found that when teens have parks nearby, they use them more frequently for physical activity, which can help lower obesity and depression rates, among other benefits. This research led to policy recommendations for decision-makers to help create access to parks and physical activity opportunities for teens. The use of geographic information systems (GIS) is a key driver in studies such as this that help direct billions of dollars in federal, state, and local funding for parks and open-space purposes that provide critical community benefits.

These benefits are often invisible or overlooked. Studies show parks provide recreation opportunities, improve mental and physical health, keep neighborhoods cooler on hot days, absorb storm water, provide access to nature and beauty, and protect wildlife habitat. Parks can improve academic outcomes and pro-environmental behaviors and create stronger connections between people and nature. The Children and Nature Network is a resource for research on these benefits. Investing in parks yields great economic returns (see chapter 6 for examples), and parks build social cohesion in communities by providing gathering places that create the opportunity for positive interactions and connection to place (figure 2.4). Parks are

Figure 2.3. In Los Angeles, California, two out of three kids do not have access to a park.

often thought of as "nice to have" amenities when they are critical infrastructure for healthy, happy communities.

In some neighborhoods, even if parks are close by, people don't feel safe visiting them, or they're not well-maintained or are too crowded. They may be surrounded by busy streets with dangerous intersections. Other parks aren't open to the public or have limited hours of operation. Park amenities might not meet the neighborhoods' needs and desires: no picnic tables in neighborhoods where multigenerational families want to gather and share a meal, no soccer fields for soccer-enthused kids, no dog parks where people can let their pets run free. Elsewhere, one park may be overused, overloved, and undermaintained, while nearby alternative parks go basically unused.

Figure 2.4. Brooklyn, New York—the New York Trust for Public Land opening ceremony of a new student-designed community playground at PS 156/392 in Brooklyn on September 20, 2019.

Parks and green spaces offer countless benefits. But for generations, low-income communities and communities of color have received a disproportionately small share of investments in parks and open spaces—and these disparities have real consequences for public health, climate resilience, and community cohesion. The causes of park inequity are vast and systemic—but its effects are measurable, and with increasingly sophisticated command of spatial data through GIS, even mappable.

These are challenging issues for park and recreation departments, "friends of" park and volunteer groups, and other organizations. GIS is critical to help manage these issues, pinpoint areas of need, and offer insights into how parks of certain types provide specific benefits.

Parks and green spaces come in all shapes and sizes: small neighborhood playgrounds, large regional parks, greenways along rivers or streams, and trails, to name a few. These varied features are opportunities to create connected, well-functioning park and green space networks within communities to serve people and nature

Figure 2.5. Locals come in droves to celebrate the opening of Olympic Sculpture Park in January 2007. The park is in Seattle's Belltown neighborhood, close to downtown. Artwork in the Olympic Sculpture Park, a new venture between The Trust for Public Land and the Seattle Art Museum, transformed a brownfield site into a world-class sculpture park in Seattle, Washington.

alike (figure 2.5). Iconic landscape architects such as Fredrick Law Olmstead and Ian McHarg studied, socialized, and popularized this concept of connection in cities from New York to Los Angeles and across the world. They drew beautiful park systems on paper—but most of their designs were never realized. Today, however, these plans are still important greenprints for what an integrated, well-functioning park system can be. It's never too late to reenvision and redesign our communities to support our quality of life and the environment.

In cities, park systems play a critical role in the green infrastructure system that enables human and natural communities to thrive. Connected park systems are havens for wildlife and enable animals—and people—to move safely and healthily through an urban environment. GIS can identify where to build new parks or restore existing sites to maximize connectivity through an entire system.

Addressing park system challenges with GIS

Data and geospatial modeling are essential tools for grappling with the scale of these park system issues. But the availability of the right data and information varies place by place, city by city. I'll provide an overview of a few standard datasets in this chapter and other datasets you'll need to seek out or create to shed light on the issues outlined earlier in the chapter. I'll cover data sources, approaches, and methods for mapping, modeling, and analyzing park system issues for your community. In this chapter, I'll focus on methods for analyzing park system access, equity, biodiversity, and connectivity. For more information on using GIS to analyze and model park systems as green infrastructure solutions for climate resilience, see chapter 3.

Gathering support for park system improvements

For these efforts, you can tap into an experienced network of organizations and agencies working hand in hand to create, design, and construct new parks, restore or renovate existing parks, create or update park master plans, engage the community, and support park operations and management. For example, the National Recreation and Parks Association, The Trust for Public Land (TPL), and the Urban Land Institute joined forces to create the 10-Minute Walk™ campaign, a national movement to improve park access and equity. Regional coalitions bring partners together across city and state lines. Chicago Wilderness uses data, science, and community engagement to advance the protection and connection of nature and communities in four states. Local organizations such as the Prospect Park Alliance in New York City focus support for local parks, plus the neighborhoods they serve.

Through existing partnerships, you'll find data, research, technical support, advocacy, and other supporting information to help frame approaches to using GIS data and analysis for action in your community. Before you get started on your project, look into what is already happening in your community and how to support, maximize, and contribute to ongoing activities.

Some park agencies have robust GIS units that provide data, tools, and applications to solve park issues, and may even make data publicly available through open data portals or hubs. But this type of GIS capacity in park and recreation departments is the exception to the rule. For the most part, park departments either have no GIS capacity or must rely on shared services from a city- or county-wide GIS department. Many park departments don't have the GIS data and analytical capacity needed to be data-driven in their daily decision-making. GIS-based analytics for park systems are usually tied to episodic master planning processes or updates. Some park departments also hire GIS consultants. When agencies and organizations have access to, and are working from, the same maps and data-driven models, it's easier to unite various decision-makers, funders, and communities behind a shared strategic park plan or vision. This approach takes data and information out of the silos within departments and organizations and makes it available and actionable for all.

Tip

Try configurable ArcGIS web apps such as Park Locator and Park Finder to create an app that provides citizens with information on where parks and recreation centers are located and information about those places.

How to approach mapping and modeling park and green space issues with GIS

The first step is to define the questions you need to answer.

Some common questions

- Where are all the parks and green spaces in the community my organization serves?
- How many people have access to these parks and green spaces within a certain distance or walk time?
- Do some populations or demographics have more access than others? You can tag individual parks with information on the population that has access within a certain distance.
- Is there an adequate number of parks and green spaces in my community, and do they have the right amenities or provide the right ecological benefits and habitat needed by wildlife?

- Are there equal opportunities for people to access park and recreation services? Does the supply meet the demand? This is often called a level of service analysis, or LOS.
- Where should parks be renovated to provide climate resilience, ecological integrity, social equity, and health benefits?

Next, determine how the results of your analysis will be used. Will it become part of a report or master plan? Will it be used to make the case for new parks in a city council meeting or news article? Is it a static printed map of park locations, or do you need an app in which users explore and interact with the data? Are you using the data to benchmark how well your community is providing parks versus another community? Understanding how the data and analytical results will be used, and the outcomes you and your partners seek, will help guide the approach you take and the products you create.

Data

Many national and global data resources include park locations, but most are incomplete, especially at the local scale. And many municipalities don't have park data digitized in a GIS format.

Where do you find park and green space data?

It's important to inspect any dataset and do spot checks to ensure the park data is complete, or have a park professional assess the data.

Start by contacting your local park department, council of governments, or regional planning organizations to see if they have a GIS layer for parks they can share. GIS data is becoming more freely available but sometimes agencies charge a minimal fee to package and deliver that data to you.

If the data doesn't exist locally, check the TPL ParkServe database, which measures park access and equity nationwide. It includes local park data for more than 14,000 cities and towns that are classified as urbanized areas by the US Census Bureau. ParkServe data is incorporated into the United States Geological Survey (USGS) Protected Areas Database of the United States (PAD-US) data product, available through the USGS Gap Analysis project web pages or through ArcGIS Online (figure 2.6). You can also download the ParkServe data from TPL. Local park data has been integrated into Esri® Vector Basemaps, so if you need only to visualize park locations, versus analyzing data, use that basemap for cartography. The USGS PAD-US database is a great resource for comprehensive US protected lands data.

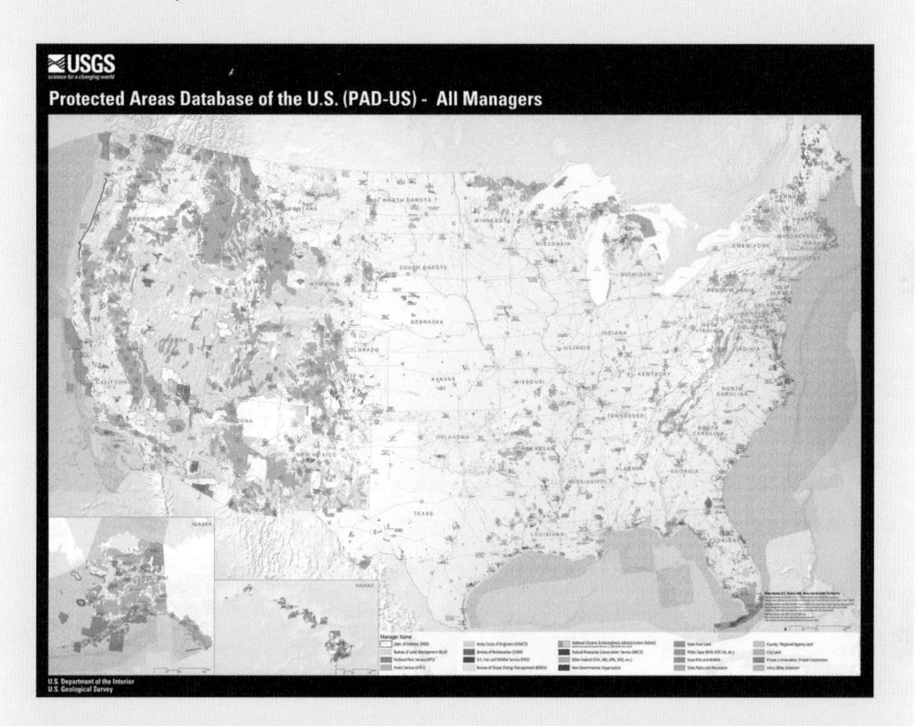

Figure 2.6. PAD-US map of protected areas symbolized according to the manager of each area.

US Geological Survey Gap Analysis Project, 2018, Protected Areas Database of the United States: US Geological Survey data release, https://doi.org/10.5066/P955KPLE

For the United States, ParkServe is a great resource, but it doesn't include every city or town in the country. If your community isn't included in the ParkServe database and you've already reached out to your local park agency, check ArcGIS Online for park data and Open Street Map (OSM). OSM is an open-source base data map that is free for all to use and contribute data to. It's especially useful when you need base data for projects working outside the United States. If you can't find park data, you may need to create the data by digitizing from hard-copy maps or using online maps and tools to locate and digitize park data. You can also collect data by directly visiting these sites and collecting data using handheld global positioning system (GPS) devices.

Tip

Be sure to check the disclaimers of all data products for potential limitations of use.

Where do you get demographic data?

Demographic data describes socioeconomic characteristics of population such as age, race, and ethnicity. This data is important in GIS analysis for conservation and parks because it tells a story about who is being impacted by land protection decisions and can provide insights on equity and environmental justice issues. This data is available from the US Census Bureau as the decennial census; the American Community Survey (ACS), a Census Bureau program; and from third-party providers, including Esri. The decennial census provides counts of population and housing that are used to direct federal funds, whereas ACS data describes changes in the socioeconomic characteristics of communities. Demographic data sources differ in their methodologies and approaches to develop the data and deliver at varying scales. It's important to know what demographic variables you want to consider and the scale (i.e., state, county, city, census tract, block group, and so on) for which you need to produce analytical results before choosing a data source.

Working with the raw data from the US Census Bureau can be complicated and time consuming but may be suitable when you need to analyze just a few variables. It can also provide more attributes, such as households without a car. But if you want to analyze many attributes across multiple geographies, which requires downloading and preparing large collections of data, consider using ArcGIS or another product that has the data aggregated and synthesized for ease of use. Census.gov has helpful "how to" resources, as do Esri Press and many universities.

Through ArcGIS Living Atlas of the World, you can access ready-to-use ACS census data to be used in ArcGIS Pro and ArcGIS Online, ArcGIS configurable mobile apps and dashboards, and ArcGIS StoryMaps stories. You will need to determine whether the attributes you need are included in the ACS dataset, or whether you'll need to pay a fee to third-party data providers to find attributes not included in the ACS product. Sometimes buying a dataset saves time and money compared with preparing the data manually depending on the scale and complexity of your project.

You may be able to find resources in which data is already collected and aggregated, such as the Environmental Protection Agency's Environmental Justice Screen, or EJSCREEN, which combines environmental and demographic data into maps and reports that are useful in park system planning and decision-making (figure 2.7).

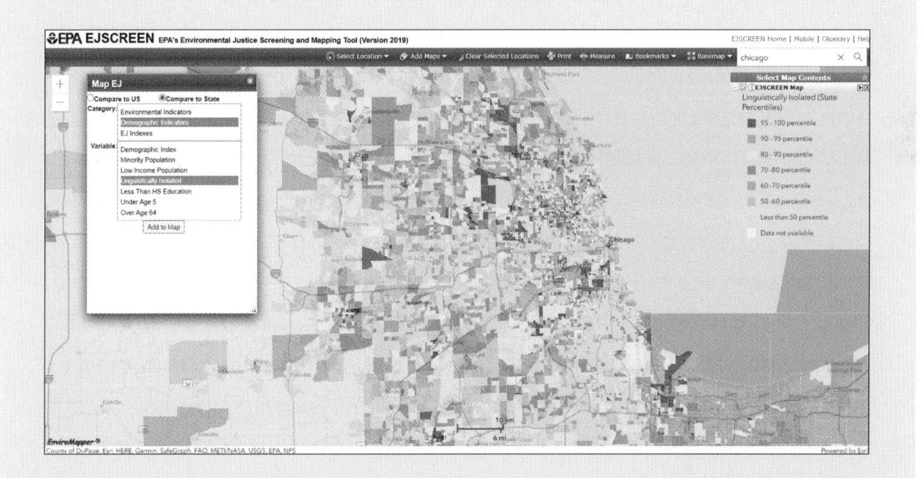

Figure 2.7. EJSCREEN interactive map shows linguistic isolation communities. Other data to explore includes demographic characteristics, along with environmental justice information such as public health and environmental indicators.

United States Environmental Protection Agency. EJSCREEN. https://www.epa.gov/ejscreen August 25, 2020.

Creating new data

The Internet of Things (IoT) sensors, video, phone apps, satellite, and other technologies are advancing quickly and can create real-time data to help park managers make informed decisions. For example, advancements in video allow park managers to combine video technology with machine learning to detect not only how many people are using a park, but a user's approximate age, the amenities they use, and how they move through the park. Anonymized cell phone data shows where people travel from to visit a park, how long they stay, and what time of day they are there. (See chapter 10 for more information on Resilient Solutions 21 and TPL's Public Land VIBE project.) This data also shows what areas of a park people don't use, which can help park managers understand how the park should be improved: Are underused areas poorly lighted, unsafe, or not easily accessible? Companies such as RESOLVE and Esri's Conservation Solutions are incorporating satellite data with artificial intelligence (AI) machine learning algorithms that detect the speed and trajectory of big-game herds on the move and link those movements to nearby poachers. Park managers deploy park security immediately, in real time, to these areas to stop poachers in their tracks. This is a promising area of data creation and analysis. See chapter 10 for more information on these and other technologies of the future.

Where do you find biodiversity data for urban or city green space planning?

Biodiversity is one of the key indicators of a healthy and thriving ecosystem. Protecting habitats that support biodiversity is one of our most powerful tools to support climate resilience and avoid ecosystem collapse. New parks, open spaces, trails, blueways (water trails for water recreation, such as boating or swimming), and greenways in the urban environment are important for creating and linking habitats that support biodiversity. Biodiversity is a key factor in landscape and regional planning and will be discussed in other chapters in the book. In this chapter, we'll focus on how biodiversity data can be integrated into urban green space maps and plans.

You can find biodiversity data from the local or state department of natural resources, the city, the county, local universities, local land trusts or conservation organizations, and the local agricultural extension office, to name a few. There are many biodiversity datasets available. Be sure to search for data that is relevant to your analysis goals and locally sourced, if possible. A few examples of biodiversity data available include ecology; biology; species occurrence and abundance; observance data from bio blitz or local crowdsourcing events; Natural Heritage Inventory data; wildlife corridors; trees, avian, riparian, and marine data; and much more, depending on the location of your city or town (figure 2.8).

Figure 2.8. Maguire primrose (*Primula maguirei*)—NatureServe global conservation status: critically imperiled (G1); ESA listing status: threatened.

Photo by Larry England, US Fish and Wildlife Service.

If your goal is to site new land protection opportunities in an urban environment that provides habitat and connectivity for wildlife, the GIS data must be at a resolution that supports site selection at that scale.

But most biodiversity data is created or aggregated at a 30-meter resolution, which potentially will not translate to accurate site selection of parcels smaller than 100 acres. Most available or vacant parcels in cities and towns are going to be smaller than 100 acres. So, 30+ meter data will reveal that a geographic area or a large parcel is potentially of value. Data at this resolution is better suited to help inform land trusts and park agencies where further investigation or site visits are needed. But that doesn't mean the right data at the right resolution doesn't exist: many universities have departments that focus on biodiversity issues and can help find the right approach to identify or create data for your project. Also consider data that could be used as bioindicators to assess environmental quality and changes over time. Bioindicators include biological processes, species, or communities. A good example of a species bioindicator is cutthroat trout. It has a temperature sensitivity to warm waters,

Figure 2.9. Florida scrub jay (*Aphelocoma coerulescens*)—NatureServe global conservation status: imperiled (G2); ESA listing status: threatened.

Photo by Mike Carlo, US Fish and Wildlife Service.

and its presence or absence can be used as an indicator for changing or warming water temperatures (Holt and Miller 2010).

NatureServe, a global organization focused on biodiversity data, created a series of datasets that power the Map of Biodiversity Importance tool or MoBi. MoBi is available through NatureServe's website and through ArcGIS Living Atlas. This data includes a rich library of species data, status (e.g., imperiled), range size, and degree of protection (figure 2.9). This data was aggregated and made publicly available, at 30-meter resolution, to support local and regional land protection planning. Figures 2.10 and 2.11 show imperiled species data for 2008 and 2020, respectively. NatureServe says the data is best used "in conjunction with field surveys and/or documented occurrence data, such as is available from the NatureServe Network." (MoBi data is important for regional and landscape planning and is referenced in other chapters.)

Figure 2.10. NatureServe Rarity-Weighted Richness Model of Imperiled Species 2008. Rarity-weighted richness of globally critically imperiled (G1) and imperiled (G2) species in the lower 48 United States, calculated using documented species occurrences from the NatureServe network (circa 2008) generalized to a hexagon grid.

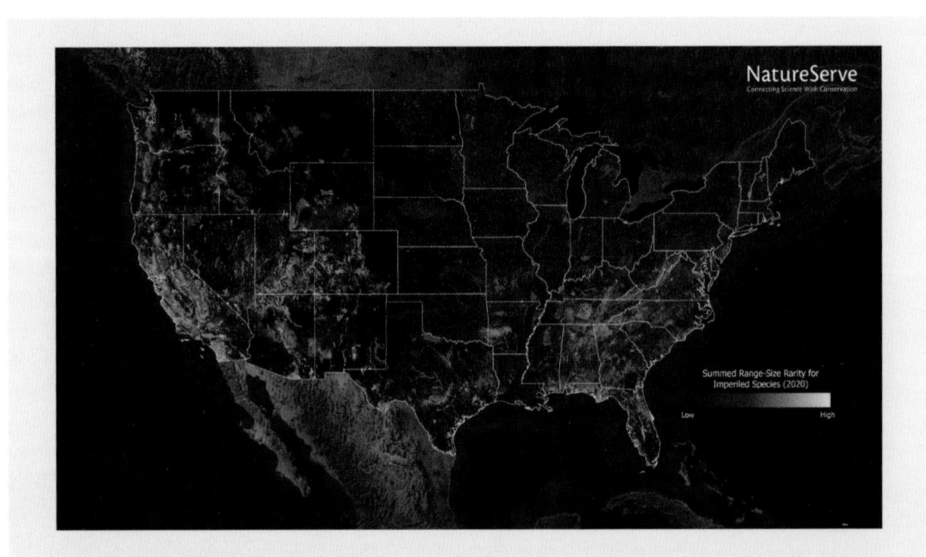

Figure 2.11. Rarity-Weighted Richness of Imperiled Species from NatureServe's *Map of Biodiversity Importance 2020*. Summed range-size rarity of imperiled species in the lower 48 United States (i.e., species that are protected by the Endangered Species Act and/or assessed as critically globally imperiled [G1] or imperiled [G2] by NatureServe), calculated using predictive models of species habitat developed for the Map of Biodiversity Importance project (2020).

Other sources include the USGS Gap Analysis Project, or GAP, which maps species range and distribution, and this data informs the MoBi data. The National Center for Ecological Analysis and Synthesis at UC Santa Barbara has also created a database of species lists, abundance, and habitat type for urban species in more than 150 cities around the world.

Where do you find data for a connectivity analysis at the city or town scale?

Connectivity has different meanings depending on the subject. For example, landscape connectivity is defined as the extent to which movements of genes, propagules (pollen and seeds), individuals, and populations are facilitated by the structure and composition of the landscape (Rudnick et al. 2012). Chapter 4 provides a deeper overview of landscape connectivity approaches and examples. Urban connectivity includes both socioeconomic and ecological characteristics operating together to provide urban system and ecosystem services. Urban connectivity can include more natural features such as riparian areas along rivers, streams, and waterways. It can also include built features, such as bike lanes and built trail systems and everything in between. A well-connected urban green space system ideally includes, among other things, parks, open spaces, trails, and natural areas that wildlife and people can use to move about the urban area through an interconnected system. The *Esri Green Infrastructure Initiative* booklet is a valuable resource that describes connectivity concepts and approaches to analysis.

Common datasets to use when analyzing urban connectivity are parks, trails, bikeways, greenways and other protected lands, riparian areas, hydrology features such as rivers and streams, satellite imagery, vacant and city- or county-owned parcels, private lands managed with a conservation intent such as conservation easement, Esri Green Infrastructure intact cores (minimally disturbed natural areas), and community forests (forests permanently protected and owned by a local government or nonprofit and managed to benefit the community), among others.

Analysis/modeling methods for park and green space systems: Park access, equity, biodiversity, and connectivity

Most conservation and park groups need to answer a baseline set of questions. This section provides an overview of select approaches and methods that the GIS analyst can use to address these common questions.

Park and green space access

Understanding where parks and green spaces are located and whom they serve is an important step in identifying park access issues. Knowing the numbers (i.e., how many people are served) is important, but understanding exactly where park gaps are located makes it possible to invest first where parks are needed most.

To generate defensible statistics, you need the right modeling approach. For instance, consider two different analysis approaches to building buffers, or "service areas," around parks to determine how many people live within walking distance.

The first is the circular buffer approach, in which you generate a circular feature around the centroid of a park using a chosen distance as a radius—say, half a mile—and then count every person who lives

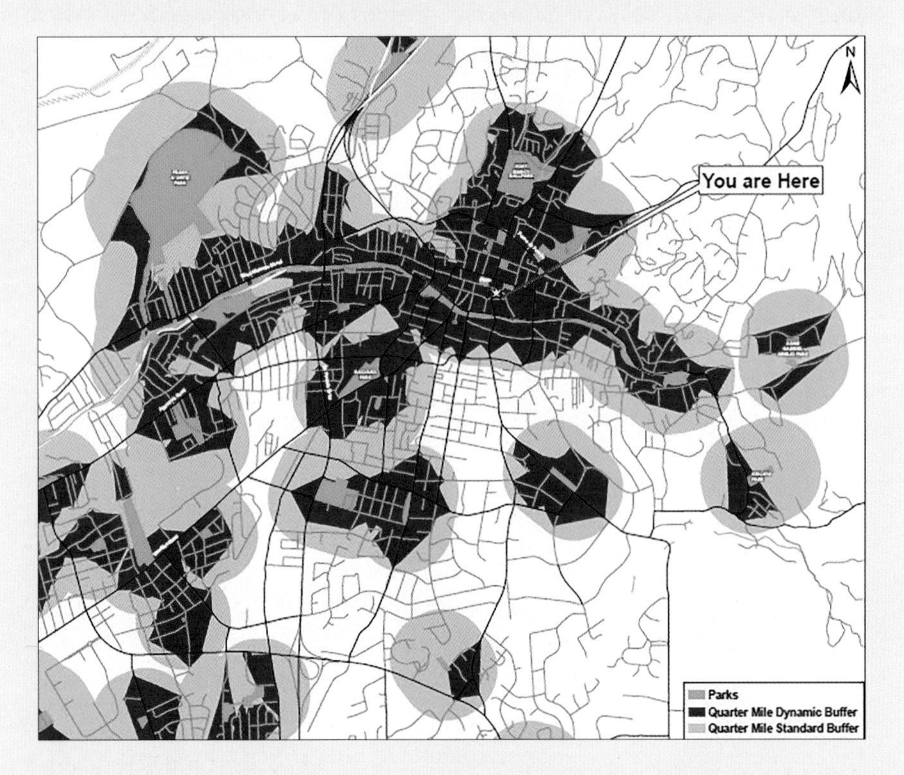

Figure 2.12. Circular buffers in orange and network buffers in red.

within that radius. The second approach uses ArcGIS® Network Analyst to develop a walkable service area using inputs such as distance (e.g., a half mile), road or sidewalk networks, or barriers such as impassable highways or water features.

In figure 2.12, you see the difference between a circular buffer and a service area approach for Santa Fe, New Mexico. In this analysis, the circular buffer analysis (in orange) returned a result of 60 percent of the population served by parks—but by not registering obstacles to walkability, this approach "pretends" that people walk to the park in a straight line, through buildings, across freeways, and over rivers. The Network Analyst service area approach, on the other hand, enables the user to include more nuanced parameters that better depict how a person might walk, drive, or bike to a park. In Santa Fe, the Network Analyst service area analysis (in red) returned a result of 31 percent served. That's a big difference. The circular buffer approach therefore overcounts the number of people who have access to a given park by including people who may be cut off by physical barriers. I recommend using the service area approach when analyzing park-by-park access, as it is a more accurate representation of park access and need across a city.

Tip

Quality assurance/quality control (QA/QC) is an important step in any analysis. Visually inspect the results to determine whether the model produced service areas for all the parks in your study area, and flag any outliers or services areas that don't look correct.

For a step-by-step tutorial on creating park service areas in ArcMap, find the ArcGIS Blog post "Measuring Access to Parks" by Rhonda Glennon.

For a step-by-step tutorial on creating service areas in ArcGIS Pro, search for the "Service Area Tutorial" in the Analysis section. This example uses fire stations but the same concepts apply to creating service areas for parks.

Park and green space equity

Once you have analyzed park and green space access, you can include demographic data layers to understand equity.

Identify the community characteristics that you want to analyze, and choose the source dataset that includes these attributes. Common variables considered for assessing a community's park need include percentage of children under age 18, population density, percent minority population, and percentage of low-income households.

Next, set up a model or workflow in GIS to use the park service areas to calculate how many people are served and not served by the park system and the demographic profile of each group. In this model, it is important to clip and normalize the demographic data to calculate more accurate statistics. For example, an intersect of the service area with census block groups and a sum of the population of those block groups would, in most cases, dramatically overcount the number of people served. Clipping and normalizing the attributes for those block groups yields more accurate results. Normalize by calculating the area percentage of each block group that is overlapped by the park service area(s) and multiply demographic statistics by that factor for an estimated result. Note that census and demographic data are estimates but these types of analysis approaches will generate defensible statistics to inform park equity issues.

Example modeling approaches to assess park needs

- Visualization: Overlay parks and the service area layers with the prioritized demographic layers to visualize where parks are needed most, based on certain demographic characteristics that are visible in the park gap areas (figure 2.13).
- Create a model that assigns relative weights to each demographic variable to reflect local demographic priorities (e.g., population density weighted higher than low-income households).
- Classify and rank demographic data to generate scores on a scale that reflects park need (e.g., 0 to 5 or 0 to 3). The scores correlate with low need (0) to high need (5) for each demographic

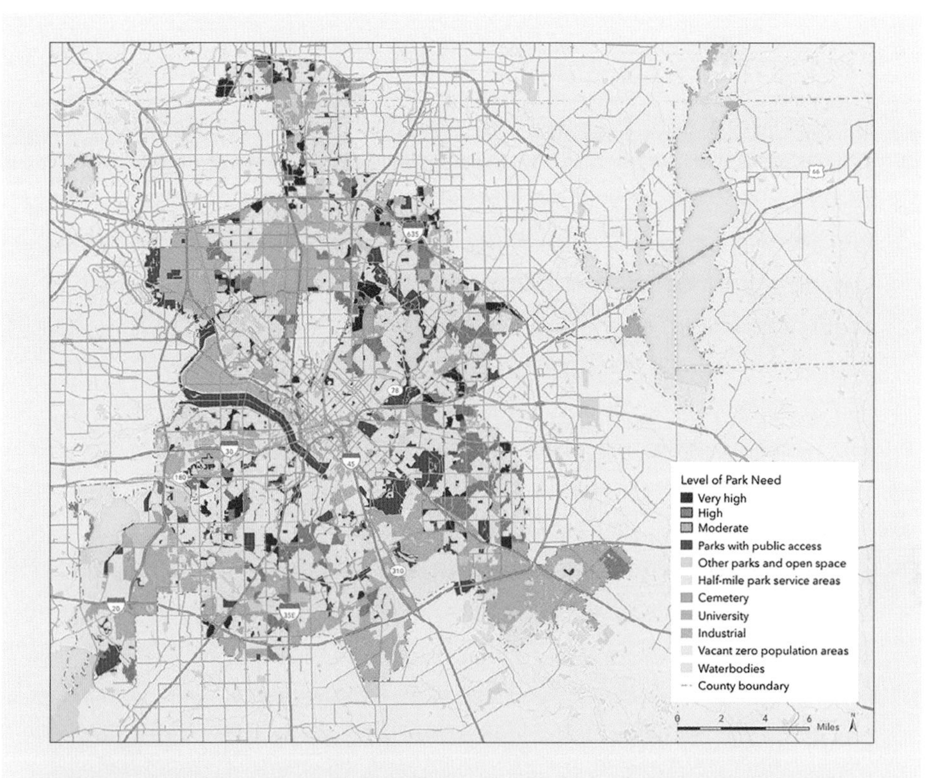

Figure 2.13. The map identifies where parks are needed most in Dallas. The red and orange areas show where people don't have access to a park within a 10-minute walk and where there is high population density, a high percentage of low-income households, and a high percentage of kids under 18 years old.

characteristic. For this method, you will need to convert the vector demographic data into raster layers to, in this case, create a ranking or prioritization that is normalized across multiple data layers.

- To generate statistics, use the calculation tools in ArcGIS Pro or ArcMap, or create a model or workflow that captures the process when you need to update statistics with new data. Creating models and workflows will enable you to create statistics for many geographies (i.e., county, city, school district, by park, census tracts, and block groups) and combine statistics for the study area.

- For QA/QC, click on individual parks and compare the number of people served against the base demographic data to spot-check results. Review the attribute table to check numbers, such as population served for a given park, and compare against demographic data.
- Using the tabular data, calculate how many people are served by parks in your study area versus those who are not. Useful tools include zonal and summary statistics. Remember to normalize by area using the method described previously. You can add this as a data point on your map or in your web app.

Define the product

A printed static map tells the story of park access visually by including base data, parks, and the park access service areas along with the demographic park need layer.

An interactive web app, on the other hand, enables more investigative power of the data. For example, users can get statistics for individual parks or for the entire park system. You can enable styling or symbolization of data in web apps to display different attributes such as park size or park manager and demographic characteristics such as age, income, and race. You may choose to include data analysis results using StoryMaps to walk the audience through a story about park access and equity in your community and why it's important to direct resources where they are needed most. These are just a few examples of how interactive web apps make your data more actionable.

At learn.arcgis.com, a tutorial on "Assessing Access to Public Transit" offers a step-by-step guide, using ArcGIS Pro, to identify areas in a city in which bus stops are needed based on socioeconomic criteria. You can adapt this approach to identify where parks are most needed based on demographic information.

Green space systems for connectivity and biodiversity

A few simple approaches to visualize connectivity issues in your city or town are as follows:

- Visually inspect the park system data (parks, trails, greenways, open spaces) with a satellite imagery basemap to identify areas that represent gaps in the park system. Sketch circles or lines on the map that show gaps and where more in-depth analysis or field observations could determine the feasibility of creating connections in those locations. Remember, connectivity doesn't always need to involve public landownership. Public/private partnerships play a big role in successful connectivity initiatives by using conservation tools such as easements and cooperative agreements, among others.
- Overlay the sketched map from the previous step with a parcel or landownership data layer. Identify parcels that could close the connectivity gaps, such as vacant lots, city- or county-owned lands, storm water management lands, utility rights-of-way, flood zones, stream corridors, potential trails or greenway easements on private properties, and lands owned by hospitals, schools, libraries, and churches. All of these are opportunities to collaborate on creating a connected park system that benefits humans and natural communities alike.

The green infrastructure framework by Esri provides a powerful set of data, tools, and approaches to aid connectivity analysis. The Green Infrastructure Collection in ArcGIS Living Atlas is a good starting place to find data to incorporate into your analysis (figure 2.14). Esri is also working on methods to create connected corridors at the local level. You can submit your data through the green infrastructure framework on esri.com, and your contributions will be included in a growing map of green infrastructure assets and priorities nationwide, and even globally in some cases.

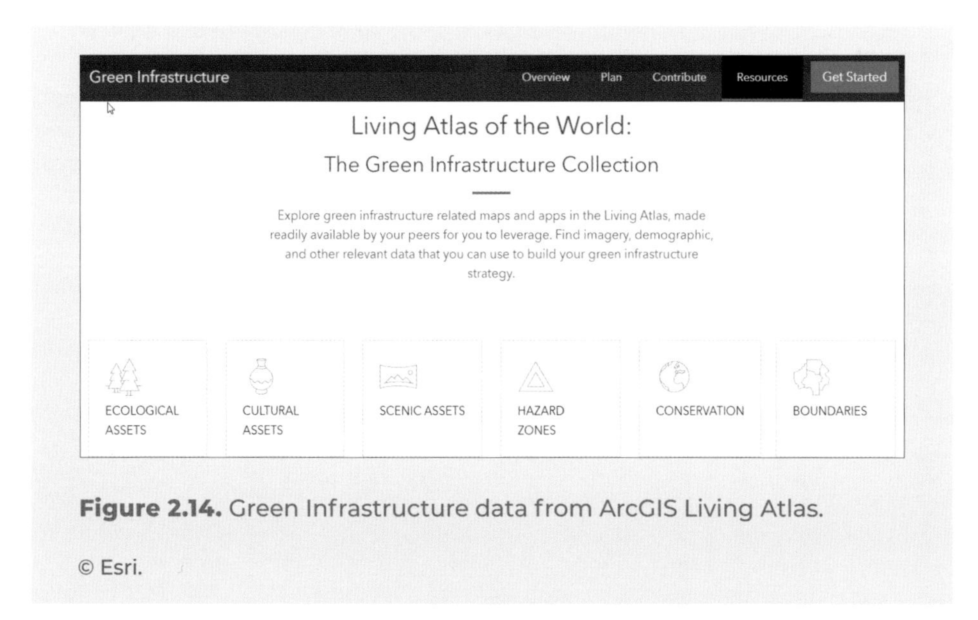

Figure 2.14. Green Infrastructure data from ArcGIS Living Atlas.

© Esri.

Translating results into action

Understanding park system issues with spatial data might be a new concept to members of your community. Visualizing the spatial distribution of parks overlaid with other GIS data is a simple technique, but it yields big insights around park equity, connectivity, green infrastructure, and biodiversity. You don't have to be a GIS technician or expert to enhance the power of data to meet the mission and goals of your organization. Simple data visualization is a powerful tool to support implementation, strategic decision-making, policy changes, and advocacy for the places we love and want to protect, restore, or enhance.

Empowering your elected officials with information on park system needs is the first step toward change. For example, knowing how many people have access to parks in your community and the demographic breakout of those communities is a first step to understanding what types of problems exist. One problem could be that children and teenagers lack access to safe parks close to home. By using demographic data to analyze where high percentages of children and teens live in relation to parks, planners and decision-makers will know where kids don't have access and start brainstorming how to address these needs. The maps may show vacant lots that could be turned into parks or school playgrounds that can be opened to the community for public use. Maps convey the spatial relationship between people and parks and help guide investments and resources (figure 2.15).

Figure 2.15. Participatory design at McKinley School in Newark, New Jersey, in 2003.

Furthermore, the ability to analyze park issues at different scales enables granular insights about where resources are needed most to address the issues. When you produce a map that shows park access by city council district in a city, such as the one for Dallas, Texas, in figure 2.16, you'll see disparities in park access by district.

These maps show where parks are lacking, and when you add more layers, they also show where park gaps overlap with other social factors, such as low-income neighborhoods or communities of color, where tree cover is needed, and where park system connectivity could benefit habitat and biodiversity. A flood zone data layer can help point to where green infrastructure can reduce the risk by absorbing storm water. These maps provide data to our elected officials so they can direct resources and funds to the places that need them most.

The more we democratize parks and green space data and make it publicly available, the more good it will do. Making data and analytical outputs available for download in a variety of configurations and file formats makes it easier for GIS practitioners and research professionals to put it to use. ArcGIS Online and ArcGIS® Hub℠ make it easy to set up and share data for your organization or agency. *ArcWatch* (2017) provides steps on how to share your data.

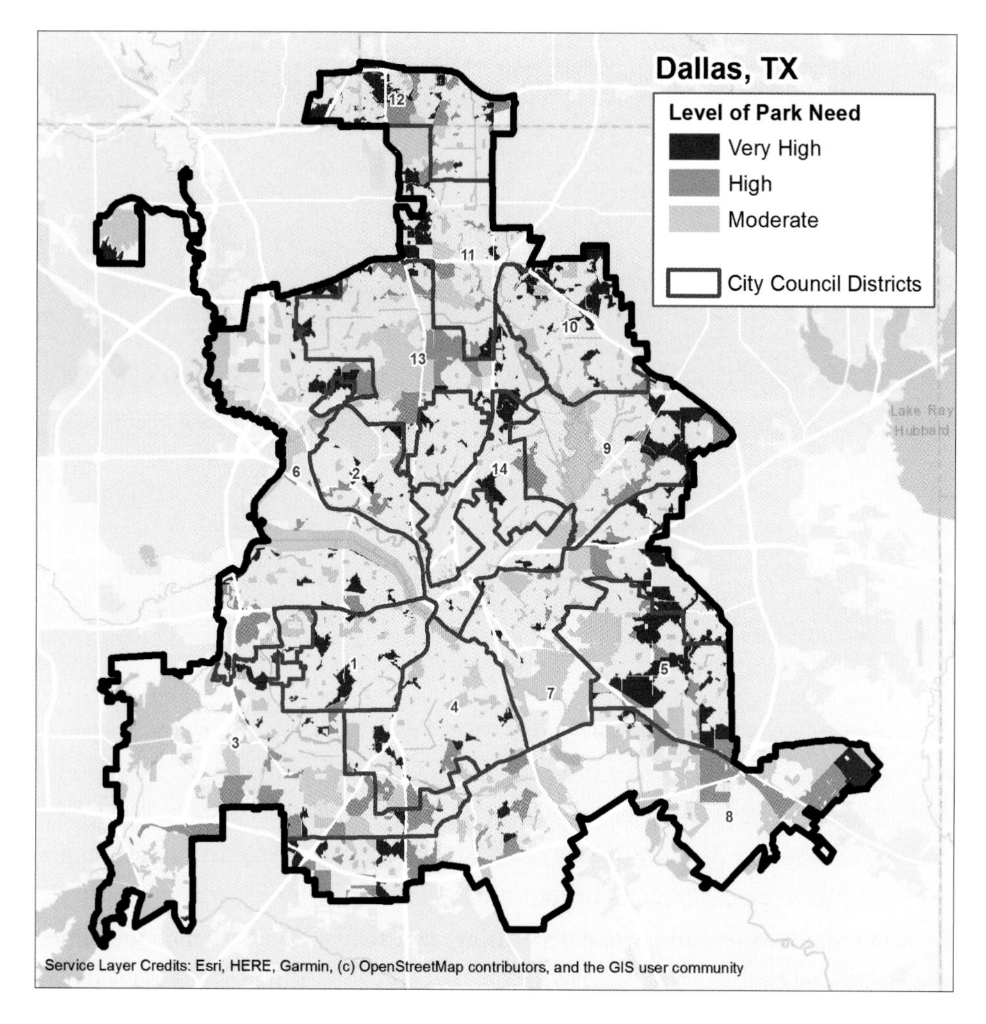

Figure 2.16. Park need areas in red, orange, and yellow.

The ParkServe platform

Created in 2018 by TPL, the ParkServe platform established the first data-driven park access baseline in the country. Now, this data is the foundation for a national movement to improve park equity.

Park planners, advocates, and citizens can access their community's park data through the ParkServe web app. Users can zoom into individual parks and find out how many people live within a 10-minute walk of that park and the population's demographic characteristics. GIS users can download the data outputs to view and analyze information at scales ranging from states to counties to zip codes.

The ParkServe platform has enabled conservation and park organizations to achieve more specificity than a national benchmark. Cities and towns are using the platform to inform the strategic siting, construction, renovation, and management of parks across the country.

Examples of how green space analysis results are valuable

Cities and towns across the country are turning to data to understand how their park systems are serving the community and nature. Many are working together, sharing methods and approaches to increase park and green space access, protect and restore nature, and address equity issues. Mayors and city officials realize that green space systems play a key role in the quality of life in their communities and are factors in attracting new residents.

Community organizing groups are using data to highlight systemic inequities and make the case for more parks and nature in neighborhoods. Organizations are developing innovative ways to increase green spaces in neighborhoods that are already, for the most part, built out. In Los Angeles, an organization called Lot to Spot is using GIS data to pinpoint where vacant or underused lots can be turned into parks, community gardens, or gathering places. In California, representatives are using park equity maps to prioritize state-level park funding for smaller communities that have the highest need.

Cities are using GIS data and analysis results to make the case for park ballot measures that provide much needed funding to build new parks, restore old parks, provide programming through parks and recreation facilities, and for operations and management. Since 1998, cities in the United States have created more than $80 billion in park funding through local ballot measures (The Trust for Public Land n.d.]. For a specific example, in 2016, TPL established a partnership with the City of Dallas to reimagine how parks, trails, and green spaces can improve equity and quality of life for the city's most vulnerable residents. Working with stakeholders, they developed a GIS-driven decision support tool—using advanced analytics to combine health, social, and environmental data with community-articulated priorities—to guide strategic green investments in the city. In 2017, Dallas city residents passed two bond measures providing $311 million in funding for parks across the city. Since then, the percentage of city residents with a park or trail within a 10-minute walk from home has increased from 58 to 71 percent, with an additional 251,000 people now served by a close-to-home park.

Organizations such as the National Recreation and Parks Association use GIS data to support advocacy, policy change, and research. Urban Land Institute uses data to find solutions to community park issues and provide technical assistance. Many universities are using park data in research ranging from park equity, climate change, urban biodiversity, park visitation, operations, and management to community health and many other topics.

How GIS improves urban green space systems and park equity

NatureServe's MoBi biodiversity data and other high-resolution sources provide invaluable information to understand where land protection in the urban realm will benefit species. Methods and approaches to mapping urban park system connectivity provide a way to connect all the green spaces into a system that benefits both people and animal species. The ability to generate walkable service areas for green spaces and parks using GIS has made for more accurate and authoritative statistics on park access and equity, more reflective of what people experience when using their park system.

Integrating demographic data such as population, age, income, and race with GIS-derived service areas is key to understanding disparities in park access, investment, and quality within and between communities. GIS is the only platform that does all this with out-of-the-box tools, and it delivers results through a variety of products, from digital maps to interactive web apps and StoryMaps stories.

With new advances in technology and IoT, park agencies and organizations can capture information on how the park system is performing to understand trends and where resources should be directed to increase, manage, or distribute green spaces for people and nature. Your analysis can support a new vision for an equitable, natural, connected park and green space system in your community and provide the data and tools needed to educate and inspire action.

References

ArcWatch. 2017. "Create an Open Data Site in Three Steps." *ArcWatch* (December).

Babey S. H., J. Wolstein, S. Krumholz, B. Robertson, and A. L. Diamant. 2013. "Physical Activity, Park Access, and Park Use among California Adolescents." Los Angeles: UCLA Center for Health Policy Research.

Holt, E. A., and S. W. Miller. 2010. "Bioindicators: Using Organisms to Measure Environmental Impacts." *Nature Education Knowledge* 3 (10): 8.

Rudnick, Deborah A., Sadie J. Ryan, Paul Beier, Samuel A. Cushman, Fred Dieffenbach, Clinton W. Epps, Leah R. Gerber, et al. 2012. "The Role of Landscape Connectivity in Planning and Implementing Conservation and Restoration Priorities." *Issues in Ecology* 16 (Fall): 1–20.

The Trust for Public Land. 2015. "Everyone Deserves a Park." Accessed January 30, 2021. https://www.tpl.org/10minutewalk.

———. n.d. "What Conservation Finance Is Really About." Accessed January 30, 2021. https://www.tpl.org/how-we-work/fund.

CHAPTER 3

Climate resilience, green infrastructure parks, and conservation

Above, a giant oak tree.

Photography by Tree Webb.

A recent survey by the Land Trust Alliance found that more than 60 percent of US land trusts are building awareness of climate change or incorporating it into their business practices in some way. Conservation organizations and land trusts are protecting forests that sequester carbon and working with energy companies to site renewable energy or transmission lines that avoid or decrease environmental and habitat destruction. Organizations such as NatureServe are creating and making available biodiversity data on species richness and imperiled habitats. The Nature Conservancy (TNC) has created data on places that will serve as important climate refugia for species. The National Audubon Society completed a study that shows where bird species might live in the future as the climate changes. Shockingly, the research findings show that, based on the pace and scale of rising temperatures, two-thirds of the bird species in North America are at risk of extinction (figure 3.1). Science tells us early interventions make a big difference in climate change, and this data will be important to help us get ahead of the impacts that places, people, and nature are already experiencing.

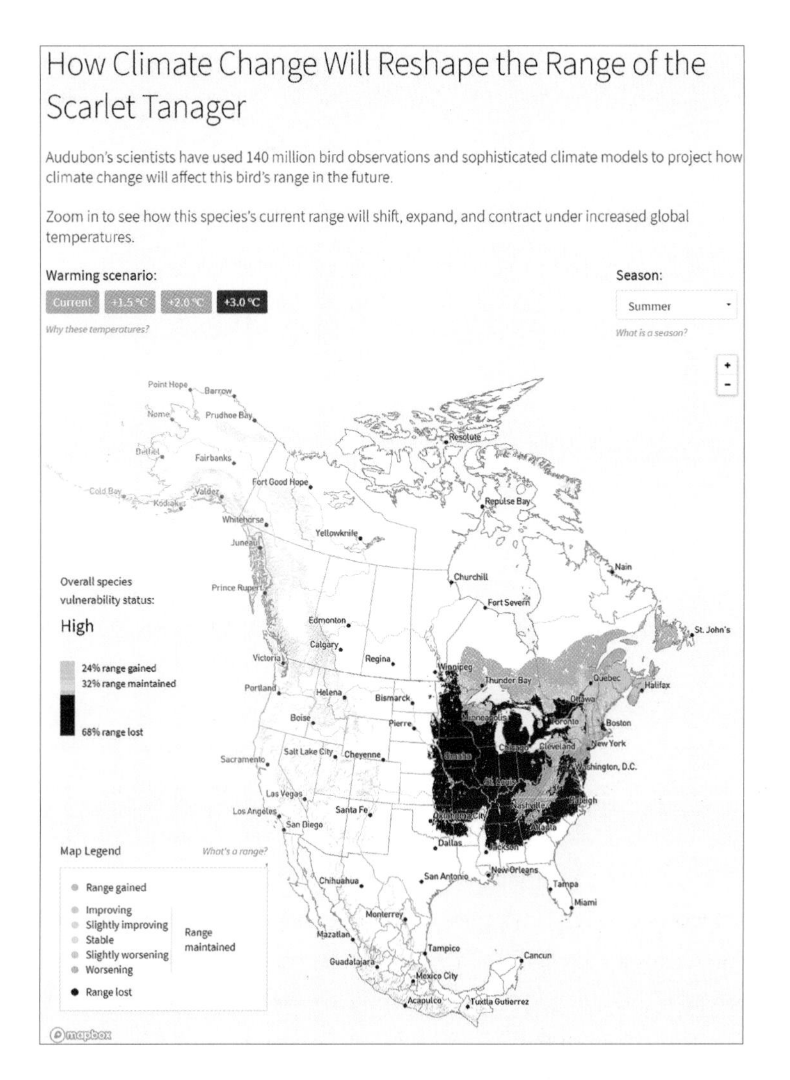

Figure 3.1. Map and information on how climate change will reshape the range of the scarlet tanager. National Audubon's Survival by Degrees: 389 Bird Species on the Brink—Birds and Climate Visualizer for Western North and Central America.

National Audubon Society.

Green infrastructure for climate resilience

Conservationists use the term *green infrastructure* to describe a network of natural areas and assets that provide environmental benefits and ecosystem services. At the community scale, park agencies and "friends of parks" groups are reimagining and redesigning parks and streets with climate resilience features that absorb storm water and keep neighborhoods cooler, among other benefits. Park systems that are designed and built using green infrastructure best practices provide many benefits, from corridors for wildlife movement to climate-resilience benefits for people and nature. Figure 3.2 shows how average global temperatures have risen since 1850.

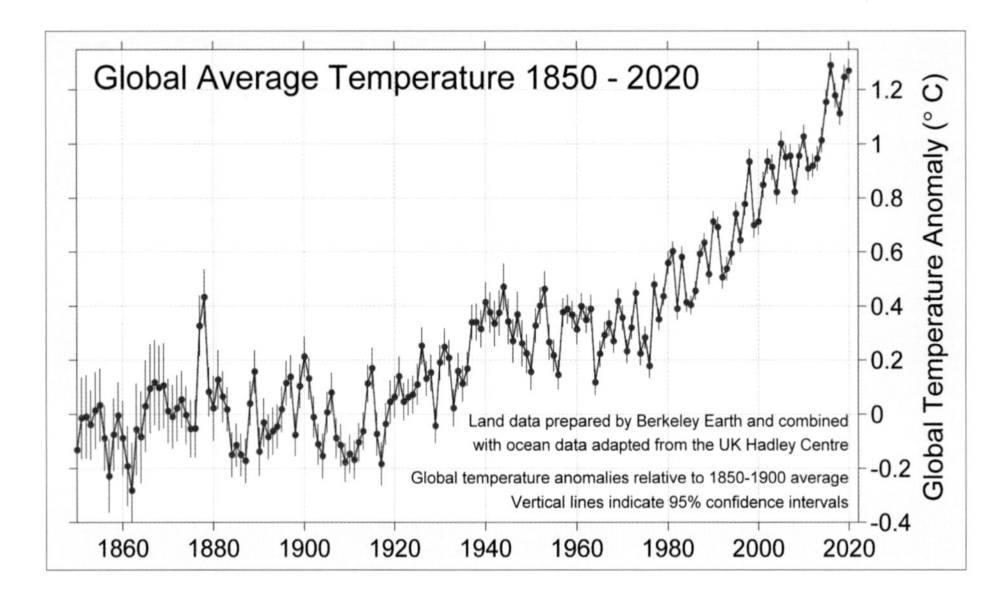

Figure 3.2. Global Average Temperature chart, 1850–2020.

Berkeley Earth. *Global Temperature Report for 2019.* Accessed 8/15/2020 from http://berkeleyearth.org/archive/2019-temperatures.

GIS is a powerful tool that helps us strategically identify where green infrastructure will provide the most benefits to communities and nature. Without it, we can't begin to understand the cause-and-effect relationships between protected lands and climate risk and change. Insights from GIS modeling and maps reveal spatial patterns in extreme weather events, frequencies of drought, and the distribution of imperiled species.

Green infrastructure

In the storm water field, green infrastructure describes capturing and treating storm water using vegetation and soil on site. Gray infrastructure is the human-engineered systems such as underground pipes and treatment facilities. For the purposes of this book, I use Esri's definition of green infrastructure and consider storm water capture and cleaning to be one of the many environmental benefits of natural, connected green infrastructure systems. For more on green infrastructure, search esri.com for the Green Infrastructure framework, or consult *Green Infrastructure: Map and Plan the Natural World with GIS* by Karen E. Firehock (Esri Press, 2019) for a detailed approach to green infrastructure methods and planning.

Climate justice and GIS

Data-driven analyses show that low-income neighborhoods, communities of color, and indigenous communities are more vulnerable to the effects of climate change and that these inequities are getting worse, not better. In cities, GIS pinpoints who has access to green spaces and who doesn't, and which communities suffer heat waves from lack of tree canopy or parks and too much pavement (urban heat islands [UHIs]). National studies have proven these inequities over and over. The maps in figure 3.3, from the American Forests organization, show clearly that neighborhoods that have high poverty, more people of color, and high unemployment are also the places that have the lowest percentage of tree canopy cover. Results from a study by Urban Studies and Planning at Portland State University, funded by the National Science Foundation and the National Oceanic and Atmospheric Administration (NOAA), show a correlation between above-average temperatures and low-income and industrialized areas of cities. The study found that in addition to tree canopy cover, a mixture of buildings of different heights, trees, and vegetation helps keep a city cool by supporting air movement and providing shade.

A lack of green space, higher levels of asthma and heart disease, and financial strain all contribute to the inequities that make low-income communities and communities of color more vulnerable to climate change. Social vulnerability data can be incorporated into analyses, used in maps (e.g., percentage of asthma by income mapped at the census block group scale), or used as overlays with climate data to show risks and opportunity areas in which green infrastructure can make a big difference (figure 3.4).

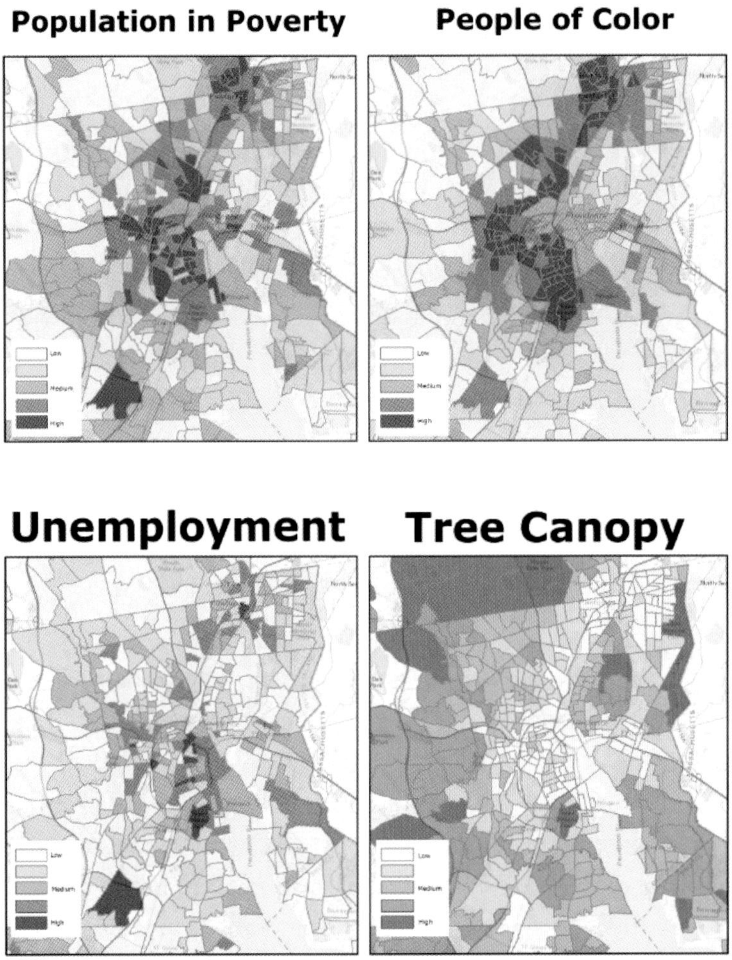

Population in Poverty

People of Color

Unemployment

Tree Canopy

Figure 3.3. Four layers of this map display percentage of people under 200 percent of the federal poverty line, percentage of people of color, unemployment rates, and the tree canopy percentage for the same study area (Providence, Rhode Island).

Rohit Musti, Chris David—American Forests, 2020.

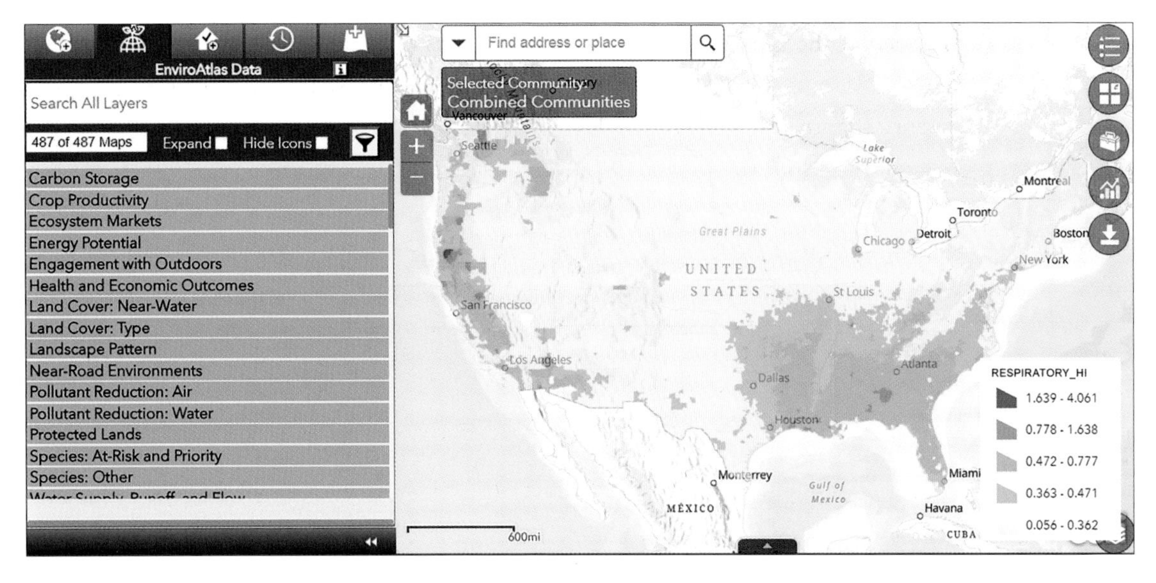

Figure 3.4. EPA EnviroAtlas interactive map showing respiratory risk (hazard index) resulting from cumulative air toxins. Yellow equals low levels of risk, and green to blue equals high levels of risk.

United States Environmental Protection Agency. EnviroAtlas Interactive Map. Accessed 8/15/2020 from https://enviroatlas.epa.gov/enviroatlas/interactivemap.

Including social vulnerability data in climate analysis

Find social vulnerability data through the Environmental Protection Agency (EPA) Environmental Justice Screen (EJSCREEN), EPA EnviroAtlas project, PolicyMap, demographic and socioeconomic data, or state- or local-level environmental justice data sources, to name a few. Some common attributes mapped for social vulnerability include low income, race and ethnicity, health disparities, social isolation, linguistic isolation, households without automobiles, and age (children and the elderly tend to be more at risk from climate issues).

The World Urban Parks organization states that by 2050, 70 percent of the world's population will live in urban areas. We need more equitable greening in these urban areas to provide healthy, livable conditions for all. A well-connected, well-designed green space system plays a critical role in the fabric of communities. Organizations are working with communities all over the world to plant and manage more trees to absorb sun and provide shade, cooling neighborhoods and villages and saving lives. They are also protecting lands that provide barriers and protections against storm surge, sea level rise, and flooding. Strong social networks are a factor in a community's response to extreme climate events. Among their other benefits, parks provide places to build these social networks and cohesion.

Local actions combine to create big impacts, so don't underestimate the power of a strategically built green infrastructure park such as a green schoolyard in New York City or a green alley in Los Angeles (figures 3.5 and 3.6). From absorbing carbon to filtering pollution from the air to providing places for social interaction and connections, parks make communities more resilient.

Figures 3.5 and 3.6. Los Angeles Green Alleys—*left*, before, and *right*, after.

Climate policy and climate data

By incrementally delivering interpreted GIS results to decision-makers, we can transform climate data into powerful decision support, implementation, policy, and storytelling products that support conservation and parks as part of the climate solution.

It helps to understand a bit about climate policy. In general, climate policy includes (1) mitigation measures to stabilize greenhouse gas (GHG) concentrations, and (2) adaptation measures that help humans and natural communities adjust to the changes that have already occurred. Conservation and park creation can be a solution for both challenges. For example, using GIS, we can identify where land protection keeps carbon sequestered in natural lands or forests or identify where new trails will provide nonmotorized commuting opportunities, resulting in lower GHG emissions. In some locations where rising temperatures are reshaping habitats, species will need to relocate to adapt to these changing habitats. If we protect the locations that species need to survive, we can minimize species loss. These are just a few examples of where GIS can address climate mitigation and adaptation issues.

The diagram in figure 3.7 shows how land trusts are responding to climate change through adaptation and mitigation actions. Mitigation actions to help reduce emissions include leading by example to reduce GHG and divest from fossil fuels. Adaptation actions that help minimize the effects of climate change include increasing buffers around waterways and increasing habitat connectivity. These are just a few ways land trusts are addressing climate change through organizational actions.

Conservation and park organizations face many climate challenges that GIS can address. I'll touch on some of these topics in this book but encourage you to seek resources focused on each of these topics and more.

No place on the planet is exempt from climate change. Around the globe, extreme flooding destroys homes, rising sea levels displace communities, heat waves are directly linked to higher mortality rates, wildfires incinerate neighborhoods and habitat, and biodiversity is disappearing more rapidly than we can measure. In the United States and globally, climate change poses the greatest threat to low-income communities that have the fewest resources to withstand these challenges. But a global movement is growing to halt or drastically slow the causes of climate change. Many are turning to science, data, and GIS for insights on climate mitigation, adaptation, and resilience.

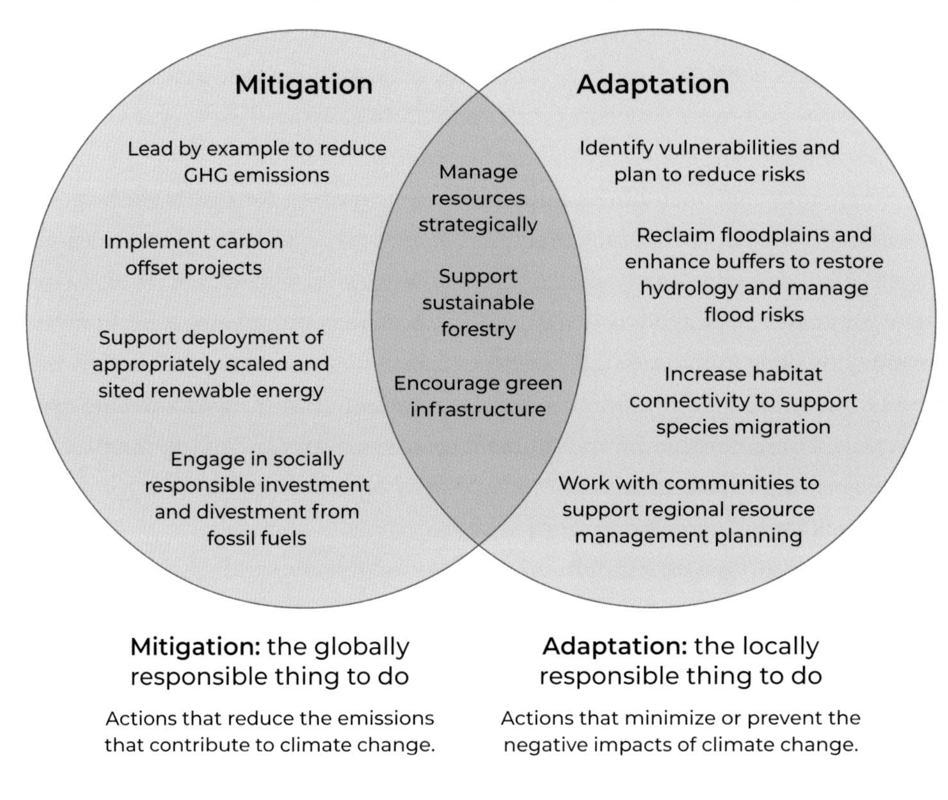

Figure 3.7. Mitigation and adaptation diagram modified by the Land Trust Alliance with permission by the Center for Clean Air Policy.

Land Trust Alliance, S. Winkelman, Center for Clean Air Policy and Green Resilience Strategies (based on J. Penney, Clean Air Partnership 2008).

More and more environmental organizations, land trusts, and park agencies are mapping and analyzing climate change to understand how to best solve or manage the challenges it presents. This hasn't always been the case. Barriers made it hard or impossible for most organizations without sophisticated GIS and research units to access the right data and science. For example, climate datasets were not freely and publicly available; were large in file size, making computation and analysis difficult or impossible for most CPUs; and were difficult to interpret and analyze without advanced knowledge. It hasn't always been easy to find and use climate-related data, but we're making progress. Many new data products are even available for free. (See chapter 8 for free online data and app resources.)

Advancements in computing power, cloud infrastructure, and the Internet of Things (IoT) are making it easier to produce data for climate modeling and tracking. Some data must be collected on-site and can be captured in real time with sensors, satellites, or other IoT methods. For example, storm water sensors can tell a park manager how much rain is falling in a park in real time and assess how well green infrastructure is absorbing that storm water (figure 3.8). Temperature sensors can pinpoint the coolest and most comfortable areas in a park, and video or sensor technology can tell a park manager how many people are visiting parks during extreme heat events. These technologies produce large volumes of data that take massive computing power to iterate through algorithms. This is where data scientists become essential. They know how to write scripts and algorithms that process data through artificial intelligence and machine learning. They have the skills to analyze large volumes of data, interpret this data, and make recommendations for decision-making. They summarize big data into more accessible and usable formats and do predictive or statistical modeling, which is essential when using climate change data to understand past, present, and future implications.

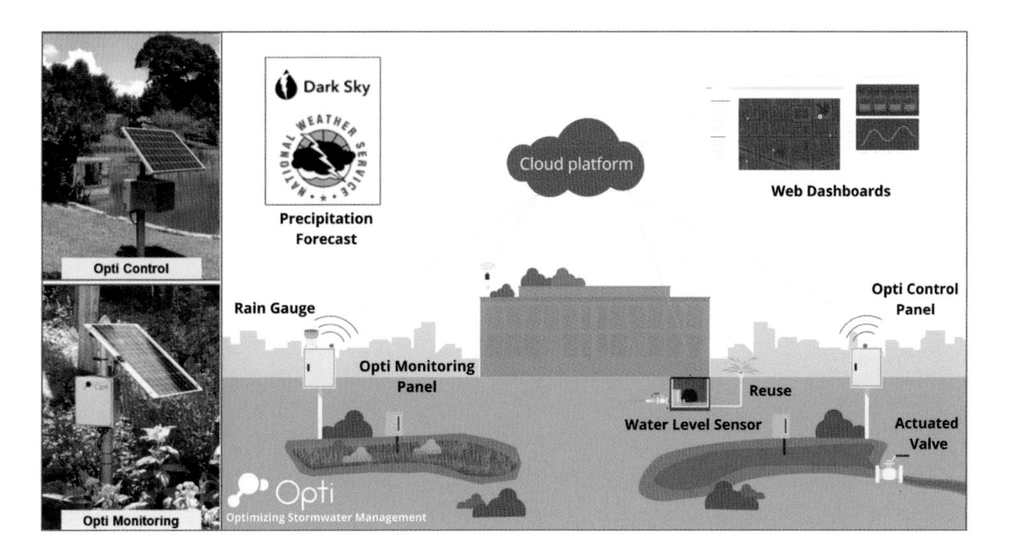

Figure 3.8. Monitoring storm water real time with the Opti storm water sensors and cloud-based platform.

Courtesy of OptiRTC, Inc. www.optirtc.com.

Strategies and methods for analyzing climate issues

Sometimes a simple overlay of key datasets uncovers exactly where we should be working to protect or restore lands for big climate resilience impacts. Other times, we need rigorous research, science, and big data analytics to support the story or narrative we want to communicate. In each case, how you interpret and present the data makes all the difference. Instead of presenting a simple map, add images and graphics to tell the story. For instance, consider creating an ArcGIS StoryMaps story that leads people through a tour of your community showing recurring flooding, including how many people are affected, the economic and social costs, and potential areas that could be protected to absorb storm water while also providing other benefits such as parkland or habitat.

The following section provides common approaches to identifying protected lands and parks that improve species and climate resilience. I'll start with common questions from land trusts and organizations and guide you through the methods for finding data and approaching the analytics. Analyzing climate issues is a vast undertaking, and the next section will only scratch the surface with three examples. The first example focuses on protecting threatened habitat for imperiled species, the second on protecting forests that sequester carbon, and the third citing where green infrastructure can address UHIs in low-income communities.

Methods for mapping and analyzing climate-related issues

Framing the question for mapping and analyzing climate-related issues

- When framing a question for a GIS analysis, consider the action and the outcome that you want to achieve with the analysis results.
- Think about your audience and the product(s) you want to create. Is the product a printed static map? A map that tells a story? An interactive web map?

- What do you want the audience to do with the information you are providing? Take action to save land? Change a policy that protects the lands you identify? Create funding to support managing or restoring lands? Inspire advocacy and support?

In the first example, the overarching goal is to protect habitat that supports endangered species and biodiversity. This goal can have different outcomes, such as protecting a certain type of habitat or preventing the degradation of natural lands. You can frame an analysis question based on the opportunity to protect lands or focus on risks or threats to the lands in need of protection. Or you can combine both.

Example 1: Protecting habitats at risk of development that support endangered and climate-imperiled species

Keeping natural lands in their natural state is crucial to the survival of species that rely on those habitats. In this example, we focus on riparian areas. Healthy, intact riparian areas—land along rivers and streams—are home to diverse plants and wildlife, filter waterways of pollutants and sediment, and provide recreation opportunities (figure 3.9).

In this example, the analysis will identify natural riparian areas that provide habitat for endangered and climate-imperiled species (the opportunity) that are threatened by development (the risk).

Your organization may be focused on a specific species or more broadly on habitat that supports many species. Once you identify what you are trying to protect, you can work with local scientists or biodiversity experts to identify potential data sources and how to use them. Many species and habitat datasets are freely available. You can do an online search for the following data sources:

- NatureServe Map of Biodiversity Importance (MoBi) data (NatureServe website and ArcGIS Living Atlas of the World)
- State Natural Heritage Inventories (e.g., Colorado Ownership, Management, and Protection database)

- For states in the western United States, the Western Association of Fish and Wildlife Agencies Crucial Habitat Assessment Tool for biodiversity and ecological data
- Audubon Important Bird Areas
- State and local department of natural resources agencies
- TNC resilient and connected landscapes data for refugia strongholds (this data will be highlighted in later chapters)
- US Fish and Wildlife Service (USFWS) Critical Habitat
- State and local fish and wildlife service departments
- Conservation Biology Institute
- Universities and research institutes

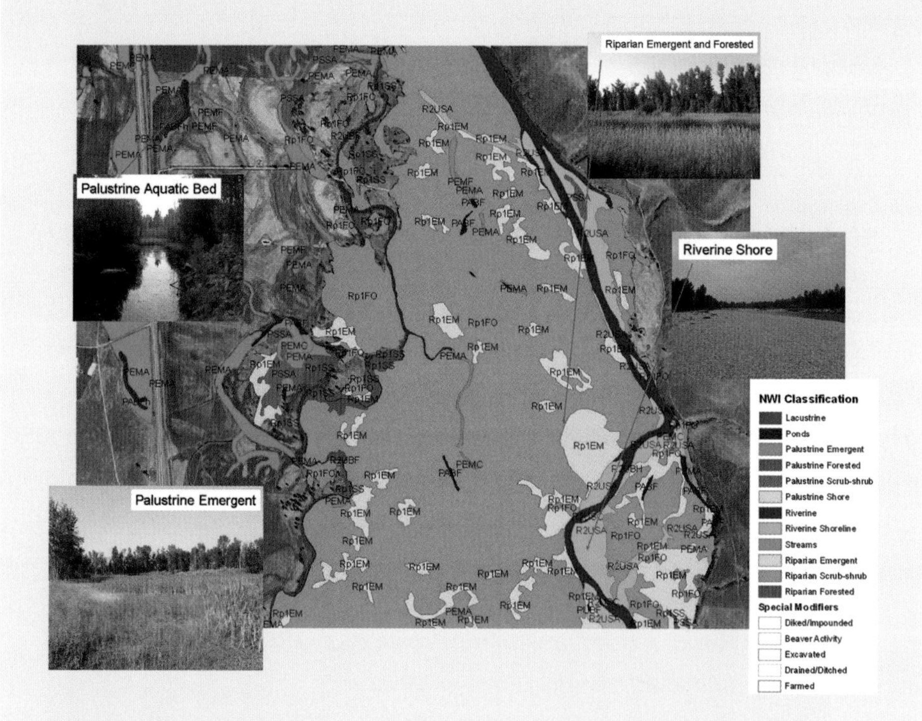

Figure 3.9. Map displaying wetlands and riparian areas mapped in Montana for the National Wetlands Inventory. Photos illustrate examples of the types of wetland and riparian areas in the map.

Sara Owen, University of Montana.

GIS analysis

Step 1

Develop the riparian areas dataset. In this example, you'll identify important riparian areas for biodiversity protection on the basis of stream buffers, wetlands, and natural land cover. In some cases, riparian areas data already exists: check with your local fish and wildlife service or department of natural resources. The National Wetlands Inventory (NWI) from USFWS is a good starting point for riparian area data in the United States.

Tip

The US Forest Service (USFS) National Riparian Areas Basemap StoryMaps story provides an overview and riparian buffer delineation models to create your own data layer.

If the riparian data layers do not exist, streams and water body GIS buffers can be a good proxy data layer. Local wildlife authorities can provide guidance on the right buffer distances to use. In some cases, simply applying a single distance to buffer stream and water body layers will create the layer you need. In other cases, you may need to apply different buffer distances to water features on the basis of stream size or order. For example, you can apply different buffers to stream segments by stream type (perennial versus intermittent) or stream order, which will be an attribute in the streams National Hydrography Dataset data layer (you can access this through ArcGIS Living Atlas).

Once created, a riparian layer can be used to focus the analysis to areas within certain distances of water features. Get more detail on habitat types and protection status within the riparian areas by overlaying data from various sources. Some examples include land-cover types from the US Geological Survey (USGS) National Land Cover Dataset (NLCD), wetlands from the NWI, or hydric soils from the Natural Resources Conservation Service. To identify areas that need to be protected for certain species, include data such as known species occurrence, nesting sites, and crowdsourced data on species sightings. This information can also help you identify lands that are already protected or are in private ownership but need restoration. Boundaries for protected areas are readily available through the Protected Areas Database of the United States or individual government agencies. Bureau of Land Management (BLM) surface ownership data can help identify public versus private lands. You can overlay data such as water quality, impervious surfaces, or degraded habitat layers to identify where habitats could be restored.

Step 2

Where are important riparian areas for species threatened by land-use conversion? From step 1, you created a riparian layer that identifies what areas are most important for protection for species. In this example, we want to prevent natural lands within the riparian areas from being converted to other uses. You'll acquire a future development layer from the local planning or zoning department. This layer will show where future development for housing, businesses, and other forms of human development will occur in the community. Overlay this layer to identify where new development overlaps or will affect the riparian area. In this example, we used land conversion or development as a threat to riparian habitat loss. But there are many other threats that contribute to habitat and species loss. In the next section, we'll explore climate-related threats and where to find data to use in your analysis.

Finding data on climate threats

It is well documented that the most common climate-related threats specific to biodiversity and habitat loss or change are the following:

- Warming temperatures
- Extreme storm events
- Drought
- Changing patterns of rainfall
- Average precipitation or temperature changes

Identify the drivers for habitat and biodiversity threats in your community, and then determine what data is available to help map, model, and analyze these threats for your organization. Following are a few of the types of threats and data needed:

- Human development (city and county planning and/or zoning departments, Theobald's development data in ArcGIS Living Atlas, census data)
- Energy development (federal agency data—especially BLM, energy company pipeline or transmission line expansion, oil and gas leases)
- Environmental threats (EPA list of impaired and threatened waters, agricultural lands expansion, general habitat loss, invasive species)

Working with data on climate threats

Visually inspect the areas that overlap, and use graphical tools to highlight areas that need protection. You can also create models that assign suitability or prioritization scores to parcels or other geographic boundary types to identify specific properties or land units that are candidates for protection. This is just one example of how data can pinpoint places to protect for biodiversity habitat goals—and land protection is just one way to protect habitat integrity, alongside land management, restoration, public education, and many other approaches.

Integrate species data into the analysis or use as overlay layers to show the areas within the riparian corridor that have occurrence, richness, or presence of species (NatureServe's MoBi data). Because riparian areas provide critical habitat for aquatic, avian, and terrestrial species, you can highlight where land-use changes would affect each one. For example, a new development may create more runoff into a stream or pollution, degrading fish habitat. A new housing development may destroy nesting or feeding habitat for birds.

Supporting imperiled species

Nature's Network is "a collaborative effort facilitated by the USFWS Science Applications program that brings together partners from 13 states, federal agencies, nongovernmental organizations (NGOs), and universities to identify the best opportunities for conserving and connecting intact habitats and ecosystems and supporting imperiled species to help ensure the future of fish and wildlife across the northeast region" (http://www.naturesnetwork.org). The collaborative created a GIS-based prioritization tool that allows you to create conservation and restoration scenarios using readily available data and methods developed by its members. This is a great example of how to use GIS to support the conservation efforts of many organizations striving to protect intact ecosystems. Figure 3.10 shows the endangered bonytail chub.

Figure 3.10. Bonytail chub (*Gila elegans*)—NatureServe global conservation status: critically imperiled (G1); ESA listing status: endangered.

Example 2: Identifying where conservation easements will have the biggest impact on conserving intact forest tracts

Intact forests are important for climate mitigation and adaptation. Sustainable forest management provides triple bottom-line benefits—social, economic, and environmental. By protecting forests, we maintain connectivity for biodiversity, provide green jobs, and store carbon. American Forests, the Longleaf Alliance, National Forest Foundation, American Forest Foundation, and others have varied data and partnerships focused on data-driven solutions for climate mitigation and adaptation. USFS is a great source of information and provides resources through national and regional outlets. Be sure to reach out to these groups before you begin your GIS analysis as they may already have the products you need.

Conservation easements

Where can conservation easements protect large tracts of intact forests that are privately owned (opportunity), and where are they at risk of being degraded by wildfire (risk)?

Potential data sources for forest data in the United States include the following:

- USFS—Forest Service Geodata Clearinghouse, USFS regional and forest resources. Forest Inventory and Analysis data, though complex, can also be used.
- ArcGIS Living Atlas—contains many authoritative datasets on forests, including fragmented forests, Green Infrastructure intact habitat cores data, and more.
- NLCD data—tree canopy.
- State department of forestry, specifically state forest action plans or department of natural resources
- Universities and research institutes

Forests are threatened by fire, human conversion and development, invasive species, and pests, to name a few. In this example, we will focus on how conservation easements on private intact forest tracts in the wildland–urban interface (WUI) could help mitigate wildfire risk (figure 3.11). Land protection is just one of many tools in the conservation toolbox that protect forests. Public-private partnerships support private landowners with large intact forest tracts to employ fire management techniques to decrease the speed at which fire might spread. Keep in mind that your analysis results can show where strategies for both public and private fire management can make a big difference in protecting large, intact forests for climate mitigation.

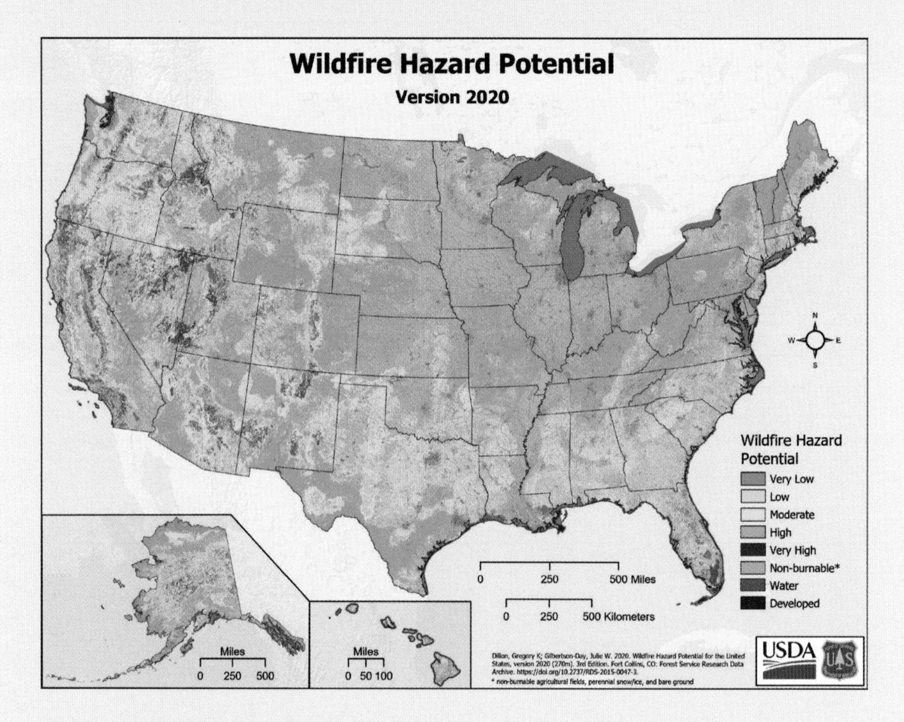

Figure 3.11. Wildfire hazard potential in the United States.

Fire, Fuel, Smoke Science Program, Rocky Mountain Research Station, US Department of Agriculture.

Data sources

WUI data sources include the following:

- USFS and local government WUI datasets
- Data.gov
- ArcGIS Living Atlas
- Parcel data—local assessor's office or national parcel data providers, such as the CoreLogic ParcelPoint® data product

Parcel data is usually available, either for free or for purchase. Advances in technology, such as ArcGIS Hub and ArcGIS® Open Data sites, and changing attitudes about open data are making it more common for cities and counties to provide their parcel (and other) data free of charge through direct downloads or ArcGIS® REST API services. In this example, you'll need an attribute that specifies public or private ownership. You'll also need an acreage field, but you can calculate that in GIS if the assessor's data doesn't include an acreage field. An attribute for vacant or number of structures on a property can also be a good indicator of suitability for a conservation easement since most limit development. The improvement value attribute or value of structures on the parcel can be a reliable way to determine whether a parcel is vacant or has structures. Because this attribute is used to determine property taxes, it is usually included and up to date. Search "improvement value = 0" to find vacant lots.

GIS analysis

To determine where conservation easements can protect and help with fire management strategies on large tracts of intact forests that are privately owned (opportunity) and where they are at risk of being degraded by wildfire (risk), you can follow this workflow.

Step 1

- Download the NLCD for tree canopy data. NLCD is a 30-meter dataset. It's best suited for identifying parcels 100 acres or larger. If you have higher resolution data for your area of interest, you can use that in this analysis.

- Download parcel data. Select parcels that are privately owned and larger than 100 acres. Overlay this selected set of parcels on the tree canopy dataset. You can visually inspect where there is a high percentage of tree canopy cover on large privately held parcels.
- In addition to visualizing the overlap, you can perform an analysis to select large, privately owned parcels that have a high percentage of tree canopy cover. One approach is to perform a "tabulate area raster" overlay analysis using the privately held parcels as the feature zone data and the tree canopy as the input class data. The results of this analysis can be joined back to the parcels to create an attribute that estimates the percentage of forest cover on each parcel.

Step 2

- With a prioritized parcel data layer based on parcel size and percent tree canopy, now determine which of the parcels are in areas with high wildfire risk. Download the Wildfire Hazard Potential (WHP), developed by the USFS and Fire Modeling Institute, or access the dataset through ArcGIS Living Atlas, ArcGIS Online, or from a local government. Use Select by Location to highlight which parcels intersect high-risk wildfire areas and export.
- Now you have a set of large parcels with high wildfire potential that could be good candidates for conservation easements. When land is protected under conservation easements, collaborative management practices and coordination with both public and private entities can lower the potential for catastrophic wildfire. Plus, the landowner receives tax benefits while maintaining the property with conservation values. See chapters 6 and 8 for details on the National Conservation Easement Database and how to use it.

Tip

For more information on the WHP dataset, see the ArcGIS blog "Wildfire Hazard Potential Enriched with Demographics now in ArcGIS Living Atlas" (September 11, 2020). The blog describes how communities are affected by wildfire and features data and ArcGIS web apps that can help communities understand risks.

The *ArcUser* article "Identifying the Most Valuable Parcels to Protect" (Fall 2019) provides the approach and methods that the Three Valley Conservation Trust uses to identify strategic land protection priorities.

Example 3: Where parks and green space are needed to help cool low-income neighborhoods as heat waves get hotter and more frequent

A study by University of Wisconsin, Madison (2019), found that the right amount of tree cover can reduce city temperatures by up to 10 degrees Fahrenheit in the summer. This is because of the UHI effect, a phenomenon in which concrete, buildings, and other elements of urban infrastructure trap and store heat from the sun, making developed areas hotter than surrounding natural areas. Tree canopy, however, creates shade, lowering surrounding temperatures. Parks, green spaces, and street trees can all keep neighborhoods cooler while also absorbing runoff during rainstorms and preventing flooding.

Analyses show that low-income communities have less tree canopy cover in cities across the United States compared with medium- and high-income communities (see figure 3.3). In this example, we'll explore how to map and analyze UHI data for your community, locate low-income neighborhoods, and determine where parks and green spaces are needed to reduce the UHI effect (figure 3.12).

Figure 3.12. Boston metro area urban heat islands.

Data sources

- Low-income data—census tracts and block groups from arcgis.com demographic data, Census.gov, American Community Survey, EPA's EJSCREEN
- Parks and open-space data—local park department, ArcGIS Living Atlas, or The Trust for Public Land's ParkServe
- UHI data—ArcGIS Living Atlas, The Trust for Public Land (TPL), or check with your local planning or GIS department or university for a possibly higher resolution analysis completed
- Parcel data—state of local planning or assessor's office or commercial source

GIS analysis

Step 1

- Identify predominantly low-income neighborhoods in your community. Census block groups are a common geography for mapping low income. Some cities' census data has an estimated average low-income household threshold. You can use this to classify low-income categories above and below the low-income definition for your community.
- Overlay the UHI data, and visually inspect where UHIs overlap with low-income census block groups. You can select low-income block groups that overlap or intersect UHIs and create a new data layer that includes low-income block groups that are in UHIs.
- Overlay parks and open-space data. Visually inspect any parks that exist in low-income UHI areas. Green infrastructure park restoration provides more greening and less impervious surface, as does planting street trees. Parking lots could be good candidates for solar panel shade.

Step 2

- Overlay parcel data and find vacant or city-owned lands adjacent to low-income housing developments. These could be good candidates for transforming into parks. Schools, libraries, and recreation centers could be candidates for tree planting or greening.
- Community organizing groups such as Lot to Spot in Los Angeles focus on converting vacant city-owned lots to parks, community gardens, and community gathering places. Working with local community organizing groups can be a great way to engage the community and understand what their needs and desires are for greening their neighborhoods. Community members are often eager to volunteer to help maintain parks and green infrastructure. This engagement creates a sense of pride and ownership in making their community beautiful and resilient.

- In addition to knowing where UHIs are located, data exists that shows future temperature projections. Explore integrating future temperature data as well as future planning and zoning data to assess and prioritize where green infrastructure can play a role in protecting people from rising heat.

There are many ways to approach using GIS to map and analyze climate change and the impacts on communities and species. These are just a few examples to get you started on framing the questions and creating the analyses that your organization is working to address. Next, we'll talk about how to interpret and use the results from geospatial analysis.

Translating GIS results into climate action or recommendations

Using GIS to prioritize places for green infrastructure parks and open spaces helps decision-makers direct funding and resources and develop new policies or update old ones. By using the best data and modeling approaches, you can defend and repeat your analysis to inform climate resilience efforts. Following are two examples, the first focusing on translating a plan into action through green infrastructure park development and the second an example of how GIS informed policy change.

The Mid-America Regional Council that serves the seven-county Kansas City, Missouri, region created the MetroGreen Action Plan in 2001 that provides park system guidance. The plan uses community engagement, GIS, public-private partnerships, and more to create a greenprint for a metropolitan trail system. The plan builds on previous work and visions in the region to connect urban and rural green corridors with the goal of protecting and improving water quality, providing recreation and commuting opportunities, and conserving and enhancing the region's existing natural places. GIS is used to identify multibenefit land protection opportunities. Of the proposed 1,144 miles of greenways, MetroGreen partners have protected more than 250 miles of greenways and almost 100,000 acres of protected stream corridors (Mid-America Regional Council 2001).

GIS informed policy change in Kinston, North Carolina, where 75 percent of the homes were damaged or flooded during repeat hurricane events. Instead of rebuilding in the floodplains, the city embarked on a community-driven planning process to relocate the homes to higher ground. GIS was a key part of the solution and was used to engage the citizens and understand their needs, as well as to map hazards, disaster response, and recovery and assess future risks. GIS was a key tool in helping the city determine acquisition and relocation and make policy decisions that have saved lives and resources and increased the quality of life of citizens (NOAA 2019).

Partnerships can fill data and knowledge gaps

TPL's Climate-Smart Cities (CSC) program focuses on helping cities identify where green infrastructure is needed to support climate resilience. The CSC framework combines partnerships, applied research, GIS and decision support apps, and green infrastructure project implementation to achieve climate resilience goals. CSC has been applied to 20-plus cities in the United States with tools, apps, and partnerships incorporated into decision-making processes. In developing the program, staff identified knowledge gaps associated with research, data creation, peer-reviewed modeling, and the interpretation of climate data for green infrastructure park issues. They began to establish a robust network of partnerships with researchers and academics, whose contributions include the following:

- Creating new GIS data at the right resolution to site green infrastructure parks in locations where they will provide the biggest climate resilience impacts
- Providing input on the features of a platform that help users easily access the data, interpret analysis results, and perform their own scenario analysis for climate resilience
- Conducting research on behavior changes that could reduce GHG emissions by getting people out of their cars and onto trails and bike/walk networks
- Improving data sources and the modeling approach for identifying where green infrastructure could reduce the UHI effect

Consider your data and methodology gaps and what local academic and research partners could contribute to improve and evolve your approach, analytics, and interpretations of data. These partnerships provide opportunities to submit research to peer-reviewed scientific journals, boost the credibility of your approach and methods, improve the science and data, and foster relationships that can be beneficial to researchers and organizations over time. Scientific and research partnerships can also boost grant and philanthropic funding opportunities in the fields of climate resilience, parks, and conservation.

How are GIS results valuable for climate resilience?

We can improve equity and climate-related social justice issues with spatial analytics. The results of GIS analyses educate city leaders about where environmental and climate issues overlap to affect the most vulnerable communities. For example, where low-income communities have less tree canopy and are much hotter than surrounding neighborhoods, decision-makers can use this data to increase green spaces and trees to lower the UHI and save lives.

With GIS, we can quantify the many benefits of green spaces and track impacts over time. This data supports continued funding for green infrastructure and makes the case for more of it across the community (figure 3.13).

How are climate outcomes different because of GIS?

GIS can help break down typical silos of government and provide data and analysis results that can be shared across departments and sectors for better and more coordinated climate resilience planning.

Maps educate local and federal officials on where conservation can aid climate resilience and guide policy development and resource allocation. GIS enables us to translate strategic climate priorities into actionable solutions that are supported by

Figure 3.13. Rainbow on a Glorieta Mesa, New Mexico, conservation property.

Photo by Melissa Houser.

science and data. Many climate resilience plans outline broad strategies but lack details on implementation. Using GIS, we can say why a specific place is important and what impacts the city and citizens can expect from protecting that place, which ensures that resources are being put to the best and highest use.

GIS also provides the framework for measuring and evaluating the climate resilience performance of green infrastructure sites in relation to the intended impacts. For example, planners can use geodesign to pinpoint where new parks are needed most and overlay climate resilience data to visualize and, in some cases, quantify the additional benefits those parks will provide for the environment and society. Geodesign, referenced in chapter 6, is defined by Esri as "an iterative design method that uses stakeholder input, geospatial modeling, impact simulations, and real-time feedback to facilitate holistic designs and smart decisions." GIS used in a geodesign process enables decision-makers to compare investments using the same data and methodologies.

GIS is a crucial tool for meeting the challenges that climate change presents to communities, species, and the natural environment. With data and analysis, we have the information we need to reimagine, revive, and reactivate the built environment and preserve natural places for climate resilience. Parks and protected lands provide many benefits to help communities battle the growing and deadly impacts of climate change. We still have a lot to learn, study, track, and evaluate regarding the benefits of protected lands in the face of climate events such as hurricanes, flooding, and heat waves. Whereas the use of GIS, data, and science was still lagging even a decade ago, these tools are foundational today in climate resilience planning and design. GIS helps us visualize and understand all the complexities related to climate and provides a powerful community engagement platform to empower locals to observe, collect, and contribute data and ideas to one of the biggest challenges we face in our lifetime. If climate isn't integrated into your vision and planning, now is the time to make it a part of your land protection strategy.

References

Mid-America Regional Council. 2001. MetroGreen® Action Plan. Accessed January 30, 2001. https://www.marc.org/Environment/MetroGreen-Parks/The-Plan/Action-Plan.

NOAA Office for Coastal Management Digital Coast. 2019. "Out of Harm's Way: Relocation Strategies to Reduce Flood Risk." https://coast.noaa.gov/digitalcoast/training/kinston-flood-risk.html.

University of Wisconsin, Madison. 2019. "Trees Are Crucial to the Future of Our Cities." ScienceDaily 25 (March). www.sciencedaily.com/releases/2019/03/190325173305.htm.

CHAPTER 4

Conserving and connecting landscapes for people and nature

Wildlife of all kinds—cougars, bears, deer, birds, bees, amphibians, and more—once moved freely along a connected ecosystem that spanned North America from Canada to Mexico along the Rocky Mountains. Beginning in the 19th century, a mass movement of settlers marched westward across the United States, and the unraveling of this corridor began. Farming, logging, mining, and roads soon began to fragment the corridors that wildlife relied on for food, water, mating and birthing areas, genetic diversity, pollination, maintaining viable populations, and much more. Today, habitat fragmentation is one of the top threats to biodiversity for our planet and for major wildlife corridors.

Above, moose crossing a wildlife overpass constructed as part of the Colorado State Highway 9 Safety Project.

Photography by Josh Richert.

Wildlands Network has developed a bold vision of reconnecting the broader Rocky Mountain corridor, which has been broken apart and consumed by human development. This impressive vision is called the Western Wildway (also known as the Spine of the Continent Initiative) and spans more than 6,000 miles from the Brooks Range in Alaska through Canada and the United States, all the way to the Sierra Madre Occidental in Mexico (figure 4.1). It includes mountain ranges, deserts, plateaus, and basins, and all the species that rely on these places. The goal of the project is to protect and restore wildlife connectivity by creating a contiguous network of conserved private and public lands. The only way to do this across such a large landscape is through partnerships and collaborations from the federal to the local level, between private and public partners, and nonprofits and corporations. To realize such an ambitious vision, the project and partnership must be guided by data.

The Spine of the Continent Initiative is an example of how local land trusts, park agencies, and conservation organizations can play a big role in landscape-scale conservation. Lands that local organizations preserve contribute to a much larger system of protected lands that provide critical habitat and ecosystem services to the communities that share that landscape. Consider the work of the Santa Fe Conservation Trust (SFCT), a small land trust operating in three counties in northern New Mexico. SFCT has been in operation since 1993 with a mission to protect culturally and environmentally significant landscapes and engage and inspire the community through volunteer opportunities and community programs. As of 2018, SFCT had protected more than 40,000 acres of land across the three counties through conservation easements, which are voluntary legal agreements between the land trust and landowners that permanently limit uses of land to protect its conservation value.

The Galisteo Basin Watershed, south of Santa Fe, New Mexico, is a focal geography for SFCT. Most of the land in the basin is privately owned. This area is critically important for wildlife habitat and movement from as far away as Mexico. It's a key passageway from the Sky Islands mountains in Mexico and Arizona north through the Sandia-Manzano Mountains to the southern terminus of the Rocky Mountain chain that leads to Canada. Golden eagles, mountain lions, bear, and many other species migrate along this corridor (figure 4.2). The basin is home to one of the largest ruins of ancestral Puebloan settlements and rock art sites in northern New Mexico. SFCT recognized the opportunity to protect this area when large ranches and planned developments began to hit the market in the early 1990s. Since then, SFCT has protected almost 13,000 acres in the watershed across 35 properties,

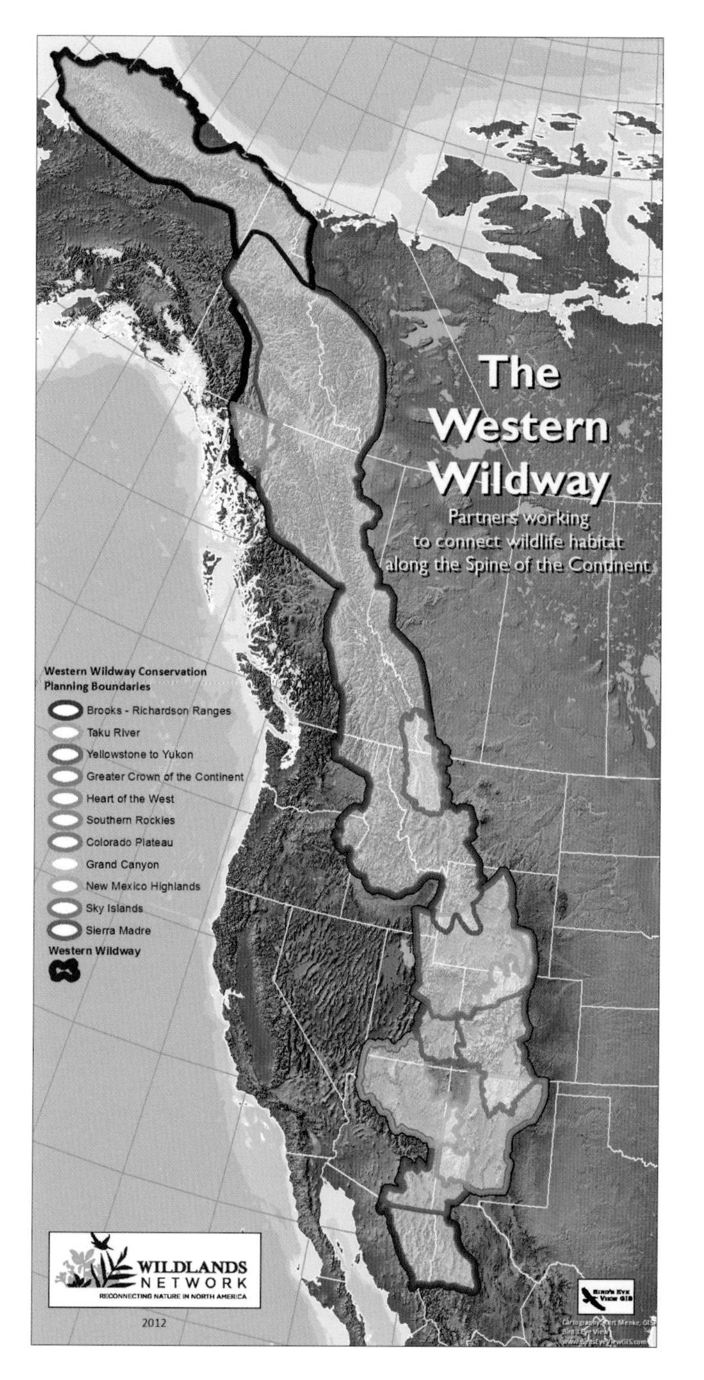

Figure 4.1. The goal of the Western Wildway is to restore wildlife connectivity along the Rocky Mountain corridor.

Wildlands Network in collaboration with Bird's Eye View GIS.

Figure 4.2. Bighorn sheep in Galisteo Basin.

Photography by Gak Stonn.

many of which are contiguous, providing important linkages for the broader Western Wildway Network (figure 4.3).

This example reminds us not to underestimate the power of local actions and conservation. These easements in the Galisteo Basin Watershed provide corridors for wildlife and waterfowl, protection for water resources, archaeological sites, and viewsheds in an area that is under fire from rapid development. There are also recreational opportunities on some of these lands for hiking and biking. These lands play a big role in protecting, connecting, and restoring a broader landscape that serves a purpose locally, nationally, and internationally for people and wildlife.

No matter your organization's geographic reach, understanding how your land protection efforts fit into a larger landscape improves your effectiveness. We can no longer justify or afford patchwork land conservation. With data, we can unite land protection efforts worldwide under master plans that protect, connect, and restore the places that support our human and natural ecosystems. Contributing to or joining landscape planning and management efforts in your area means that the goals and values of your organization are reflected in land protection actions at scale, and your organization will benefit from access to information, data, funding, and partnerships.

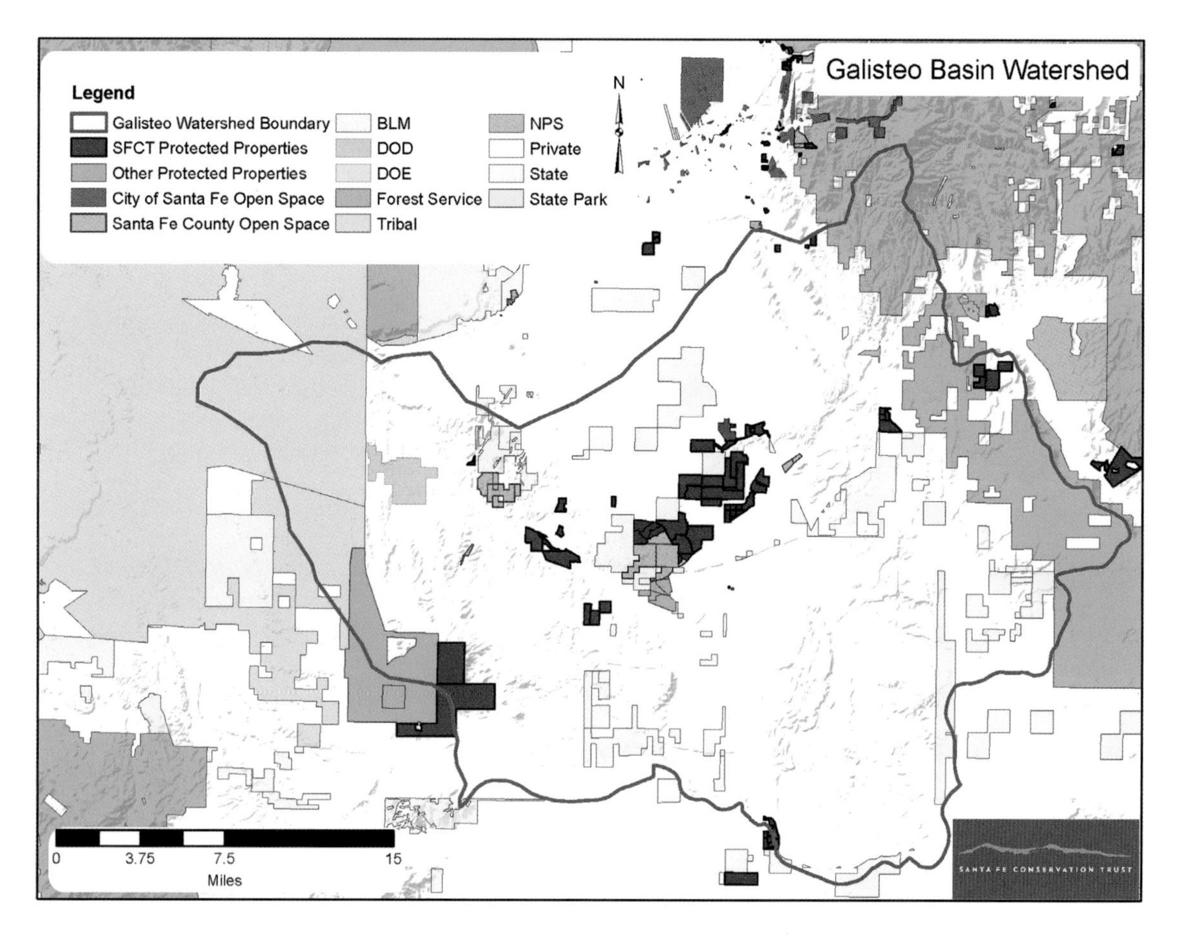

Figure 4.3. Map of Galisteo Basin Watershed protected properties, *in purple*, that provide the linkages in the Western Wildway corridor.

Courtesy of Santa Fe Conservation Trust.

The economic angle of landscape conservation and connectivity

Besides helping ecosystems, landscape conservation and connectivity makes economic sense, too. The *Landscape Connectivity Call to Action* report from the World Business Council for Sustainable Development highlights in detail the triple bottom line benefits of connectivity and corridors. The triple bottom line is an expanded way to operate a business or nonprofit by striving to achieve benefits socially, economically, and environmentally. The report gives examples of the direct and indirect benefits of connected green infrastructure. In one example, property values of homes in Colorado experience a 32 percent increase when adjacent to wildlife corridors, and crop yields increase significantly when just 3 percent of the land is purposed for wildlife habitat. The report includes a wealth of information on ideas for strengthening cross-sector collaborations, policy, and new design ideas that support the triple bottom line.

In another example, the Native Plant Conservation Campaign states, "For the US alone, pollination of agricultural crops by wild pollinators has an estimated value of $4.1 to $6.7 billion per year" (Wildland Ecosystem Services n.d.). Wild pollinators must have protected, connected habitat to survive, thrive, and pollinate the crops that sustain the world. Land protection and connectivity not only improve habitat for wildlife, they provide key environmental, social, and economic returns that help our communities thrive and prosper sustainably.

What is landscape conservation?

According to the Network for Landscape Conservation (n.d.), "Landscape conservation is an approach that brings people together across geographies, sectors, and cultures to collaborate on conserving important landscapes and the myriad ecological, cultural, and economic benefits they provide." The practice is also referred to as "large landscape conservation" or "landscape-scale conservation."

Common themes across landscape conservation definitions

- Collaboration among many partners
- Developing and using data, maps, GIS, and spatial decision support tools
- Developing policy, planning, and management strategies that attempt to balance the human and natural uses of the landscape

- Creating funding through conservation finance, dedicated US national sources such as the Land and Water Conservation Fund, collaborative programs that foundations support, and impact investments

How do partners in landscape conservation collaborations approach understanding these themes and outcomes? How does each partner know how their land protection efforts currently fit or could fit into a bigger picture? In these collaborative efforts, GIS is a critical tool. GIS helps identify what to protect and why, and provides insights into how to manage for sustainability. For landscape conservation collaborations, planning, mapping, and modeling are central to the process of producing a plan of action to guide land protection efforts.

Creating a landscape conservation plan

The steps to produce a landscape conservation plan commonly include the following:

- Identify a shared vision for the plan. Determine the stakeholders who will contribute to the vision. For example, natural resource agencies, city and county planners, private landowners, environmental NGOs, industry, researchers, and more.
- Convene the stakeholders and identify the vision.
- Based on that vision, identify what to map, model, and analyze (including but not limited to biodiversity, natural resources, connectivity, climate resilience, cultural resources, socioeconomics, and threats).
- Identify available data—consider accuracy, scale, relevance, how current the data is (vintage), and format.
- Create a landscape conservation model using systematic conservation planning tools such as Zonation and Marxan or approaches such as the methods presented in the Esri® Green Infrastructure Initiative and produce maps that guide recommendations and action.

- Identify or create funding for actions.
- Create the plan.
- Implement the plan.

References for creating a conservation plan

Ahern, J. 1999. "Spatial Concepts, Planning Strategies, and Future Scenarios: A Framework Method for Integrating Landscape Ecology and Landscape Planning." In *Landscape Ecological Analysis*, edited by J. M. Klopatek and R. H. Gardner. New York: Springer. https://doi.org/10.1007/978-1-4612-0529-6_10.

Cayton, Heather. 2020. "Technical Guides." *Conservation Corridor* (January 20). https://conservationcorridor.org/technical-guides.

Firehock, Karen E., and R. Andrew Walker. 2019. *Green Infrastructure: Map and Plan the Natural World with GIS*. Redlands, CA: Esri Press.

Landscape Conservation Network. 2018a. "Recommended Practices for Landscape Conservation Design." Version 1.0, September. https://lccnetwork.org/resource/recommended-practices-landscape-conservation-design.

———. 2018b. "Pathways Forward: Progress and Priorities in Landscape Conservation." https://landscapeconservation.org/wp-content/uploads/2018/08/Pathways-Forward_2018_NLC.pdf.

———. 2020. "Landscape Conservation Design." https://lccnetwork.org/issue/landscape-conservation-planning-and-design.

The resulting plan is a strategic road map that prioritizes large landscape conservation. Ideally, this road map is a living decision support tool to guide partners in their daily work. Making the data and products available through web-based decision support apps means everyone is working toward shared goals and understands how individual efforts fit into a bigger picture. The app can also be updated when data changes, new priorities emerge, and when lands are protected. Later in the chapter, we'll explore what it means to find and use existing data or contribute information using GIS in these efforts.

Landscape conservation progress so far

As of August 2020, 15 percent of the world's land is protected, according to the UN Environment Programme World Conservation Monitoring Centre (UNEP-WCMC), the International Union for Conservation of Nature (IUCN), and the National Geodetic Survey (NGS) (UNEP-WCMC, IUCN, and NGS 2021). For example, according to the *Protected Planet Report 2018* by UNEP-WCMC and IUCN, Latin America and the Caribbean have the highest percentage of protected lands compared with other countries. The map in figure 4.4 shows the percentage of lands protected by country from the current World Protected Areas Database.

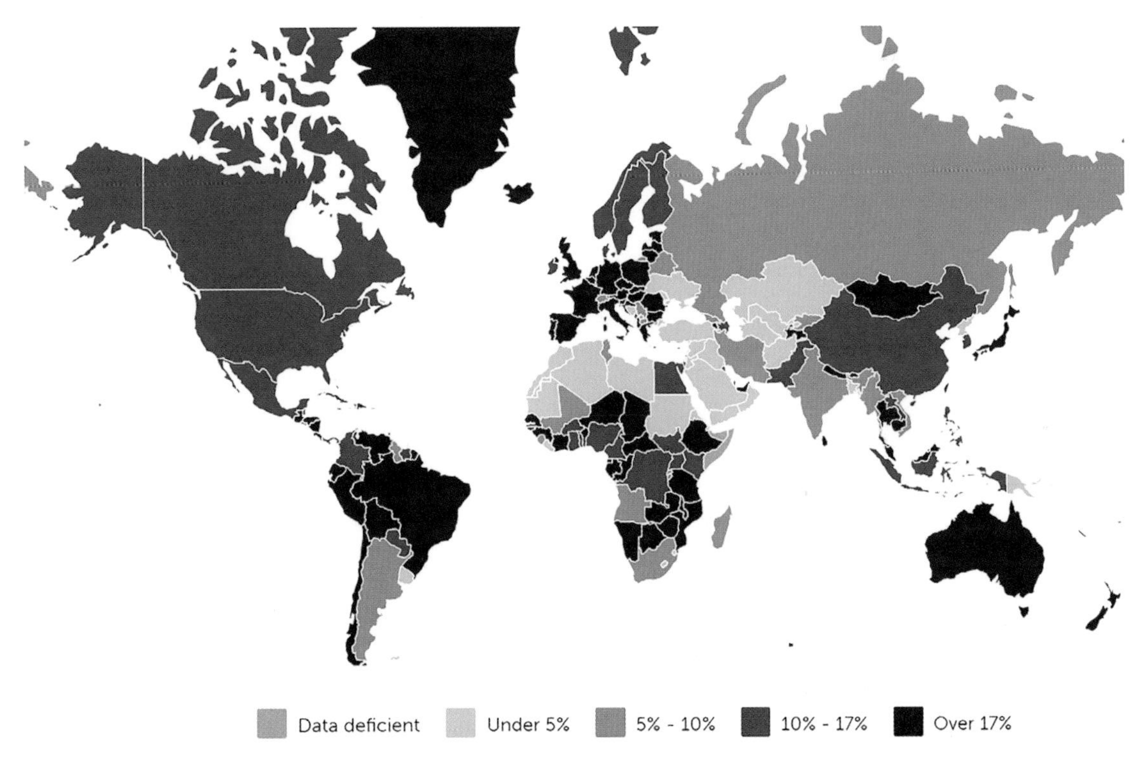

Figure 4.4. World Protected Areas Database—land protection percentage per country.

UN Environment Programme World Conservation Monitoring Centre (UNEP-WCMC).

The United States is roughly 13 percent protected. The Convention on Biological Diversity set a global goal to have 17 percent of the world's terrestrial lands protected by 2020. The United States is slightly below this goal, but many argue that 17 percent is hardly enough of protected lands, especially considering that some of the big visions, such as E. O. Wilson's Half-Earth and the Udall-Bennet Thirty by Thirty Resolution to Save Nature by US Senators Tom Udall and Michael Bennet, are calling for much more to be protected. Even if a place has a protected designation on the map, it doesn't necessarily mean that protections are a reality. For example, there are many "paper parks," places that appear to be protected by legislation but the activities and management of the areas do not offer the intended protections. Many marine-protected areas and many protected areas in countries such as Africa and Latin America lack effective management strategies to support protections.

Even though countries have high percentages of designated protected lands and waters, the management mechanisms do not provide the level of protection needed to maintain protection benefits. For example, some protected lands might allow for oil and gas drilling, unsustainable fishing practices, or other land uses that negatively affect the lands, waters, people, flora, and fauna (MPA News 2001). Many protected areas are small and isolated and can offer more biodiversity benefits when connected to a larger system of protected lands. The configuration of protected areas matters. We need many more protected and connected places to support a sustainable and resilient world.

GIS and landscape conservation

Big visions such as Half-Earth (figure 4.5) and Thirty by Thirty provide calls to action and a North Star for protecting the lands with the most biodiversity. GIS maps show us exactly where our conservation and park work will make the biggest impacts. Visions such as Wilson's owe much to GIS, which provides the data and maps to support science and implementation strategies.

Considering the threats and dangers facing our natural world, we must increase the rate and pace of land protection now. If we are going to save the most important natural places, meet the targets, and fulfill the visions set forth by groups from the United Nations to the local land trust or park agencies, we must act now and scale up our collective impacts.

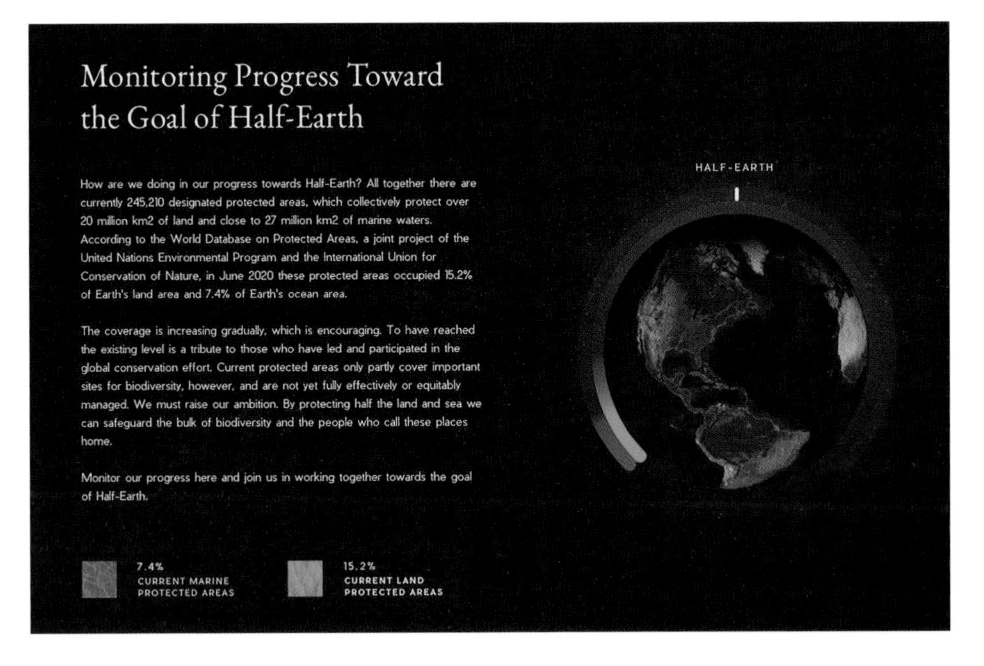

Figure 4.5. The Half-Earth website is tracking progress on land and marine protection efforts. Currently, 15.2% of the earth's land area and 7.4% of its ocean area are protected. Research shows that we must protect at least half of land and ocean resources to avert major climate disruptions and biodiversity loss.

E. O. Wilson Biodiversity Foundation's Half-Earth Project.

Connectivity and corridors

Considering that habitat fragmentation is one of the top global threats to biodiversity, mapping connectivity and corridors should be included in landscape planning processes. Existing barriers make it hard to understand how to incorporate connectivity into a plan. For example, there continues to be debate in the field around definitions and analytical approaches to connectivity and corridors. The Center for Large Landscape Conservation (CLLC) and the Wild Foundation produced a report, *Habitat Corridors and Landscape Connectivity: Clarifying the Terminology*, in which the authors (Meiklejohn et al. 2010) conducted a literature review of historical and modern definitions and synthesized the results. They define connectivity as either functional or structural. From the paper:

> "Functional connectivity describes the degree to which landscapes actually facilitate or impede the movement of organisms and processes.

> "Structural connectivity refers to the physical relationship between landscape elements."

In other words, structural connectivity refers to the parts of the landscape that are physically connected, such as wildlife corridors. Functional connectivity refers to species-specific interactions with the landscape, such as active and passive dispersal. An example of active dispersal is when animals or birds move through different geographic areas to find food, mate, or migrate. A passive dispersal example is when the wind carries seeds to another location. Landscape characteristics have a big impact on both types of dispersal. For example, animals need connected, barrier-free corridors for the easiest passage from one location to another. Human development such as highways and housing growth cuts into wildlife corridors, making it harder or creating more resistance for animals to safely move from place to place.

The authors describe habitat corridors as one of the main tools for achieving functional connectivity. Corridors support both movement and ecological processes, do not have to be linear, and can exist in different scales for specific processes. Translating this to models or algorithms is not an easy task. A 2014 study from the US Forest Service (USFS) Rocky Mountain Research Station (Baldwin et al. 2014) described landscape connectivity tools as "challenging to use" and "highly technical." Regardless, performing these analyses is critical for strategic conservation. Later in the chapter, I'll provide an overview of some of the connectivity and corridor analysis tools.

Luckily, many organizations have either completed connectivity analyses for landscape efforts or have the staff expertise to help you do the same. One of the benefits of working with a landscape planning cooperative or group is the ability to collectively identify an approach for connectivity analysis, which means the results are more likely to be used by partners at all scales to protect the corridors and lands that are most urgent and strategic.

Landscapes may cross many political, geographic, and cultural interfaces that involve complicated aspects to consider. Public and private land managers, indigenous people, policy makers, governments, and physical and political barriers must all be taken into consideration in a landscape connectivity analysis. Different approaches to education, implementation, and policy will likely be needed to achieve connectivity across these administrative, cultural, and geographic interfaces. For example, most of the world's indigenous communities live in some of the largest and last remaining natural areas. Working with these communities to support their stewardship of biodiversity and conservation is key to successful and holistic conservation efforts. Maps that tell stories, GIS decision support applications, video, and infographics can help with these efforts and provide the necessary engagement and inclusiveness.

Organizations and Agencies Working on Landscape Conservation, Connectivity, and Corridors

The Center for Large Landscape Conservation aids landscape conservation and connectivity practitioners through science and coordination. The CLLC resource library includes tools, case studies, and more. CLLC's project work includes advancing connectivity across the US national forests and working with the Blackfeet Nation's Fish and Wildlife Department in northwest Montana, for example, to reduce animal-vehicle collisions through measures such as safe road crossings, including wildlife underpasses (figure 4.6).

Figure 4.6. Buck mule deer using highway underpass in Colorado.

© Courtesy of the Colorado Department of Transportation.

As mentioned earlier, the Wildlands Network focuses on reconnecting, restoring, and rewilding North America through large, connected corridors that span from Canada to Mexico (figure 4.7). They use data, science, research, and collaborations to guide and focus conservation efforts on the places that are most important to biodiversity.

Figure 4.7. *North American Wildways* map.

Wildlands Network.

Beyond the United States, Elephants Without Borders (EWB) considers elephants to be the iconic flagship species that can raise awareness and support for conservation. Supporting elephants and their habitats will benefit other species, including humans, as conservation supports sustainable development, tourism, and collaboration between villages and countries within the region. Yet elephants experience major barriers for survival and movement in the region. Some of the threats include poaching, human development, and buried land mines that maim or kill. To raise awareness of the issues and promote land protection practices, EWB uses a landscape conservation approach in a region that spans five countries in the southern Africa region. The area, called the Kavango-Zambezi Transfrontier Conservation Area, includes Botswana, Namibia, Angola, Zambia, and Zimbabwe. By combining science, data, research, and education, EWB prioritizes, tracks, and monitors the progress of elephant recovery in the region and supports ongoing community collaboration (figure 4.8).

How to optimize land protection for corridor design by balancing multiple factors

Land protection organizations are consistently faced with the reality that there is a lot of land to protect but limited funding. How do you decide where to focus? What places are critical for optimal outcomes? Which are the most important and why? Scientists, researchers, conservation biologists, and others have studied optimization criteria and the effects of choosing one area or corridor over another. For example, when land protection goals are focused on one keystone species—species, such as wolves and beavers, that maintain the biodiversity and function of an ecosystem—the lands that need to be protected tend to be expensive because of their pristine nature. Cost can become a big barrier to protecting or restoring lands for connectivity.

Studies have found that by designing corridors for multiple species versus single species while considering the trade-offs between ecological and cost factors, practitioners will have more land protection options that are less expensive and still provide corridor connectivity. Multispecies corridor design has drawbacks, including that such corridors can perform worse than corridors designed for individual species. Multispecies corridors are generally a compromise because of limited conservation resources. So how can GIS help? Data, maps, and models are key components to support this type of decision-making by providing land protection scenarios in visual and interactive maps and apps. By understanding the trade-offs, conservationists can make better decisions.

The following sections provide examples of how GIS supports optimization models.

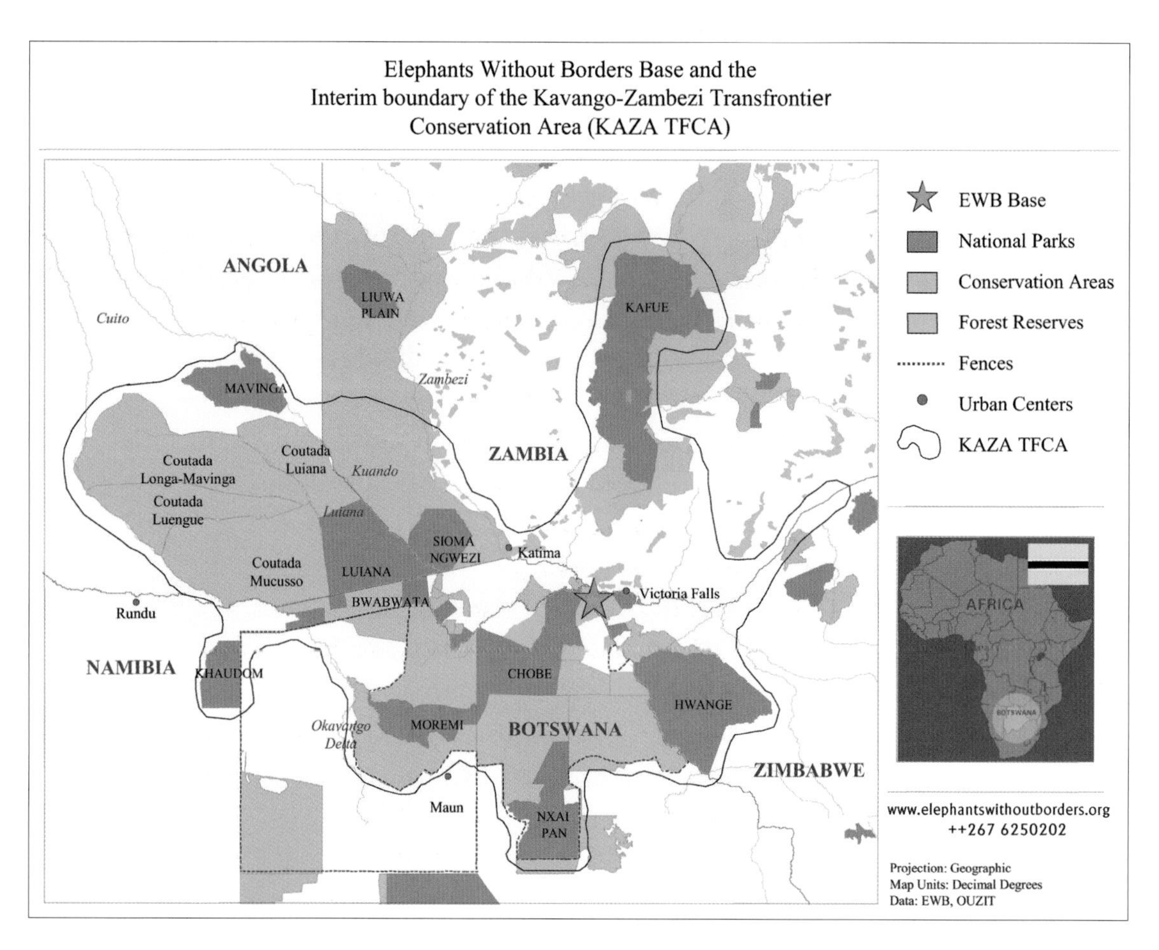

Figure 4.8. This map shows the Kavango-Zambezi Transfrontier Conservation Area where Elephants Without Borders (EWB) operates. EWB uses aerial surveys and satellite monitoring to understand populations, movements, barriers to movement, threats, habitat needs, social organization, and many other aspects of wildlife in the region. The data and maps provide sound information to identify new lands to protect for landscape conservation and migration corridors.

Dr Michael Chase, Elephants Without Borders, for conservation and management of healthy populations.

Multispecies corridor modeling in Argentina

In their research article "Using Niche Modeling and Species-Specific Cost Analyses to Determine a Multispecies Corridor in a Fragmented Landscape," DeMatteo et al. (2017) sought to identify corridors that need to be protected to restore connectivity and reduce fragmentation for five carnivores. The study took place in Misiones, Argentina, where the largest remaining tract of Upper Paraná Atlantic Forest ecoregion remains. Human development in the region in the form of plantations, agriculture, and pastures has fragmented habitat and left only 50 percent of the forest protected. The authors modeled habitat suitability, use, and species richness and performed a cost analysis for jaguars, pumas, ocelots, oncillas, and bush dog to identify primary and secondary corridors (figure 4.9). This study addressed the connectivity needs of multiple species while considering resource and cost parameters, resulting in newly defined corridors and comprehensive conservation strategies.

Large predator corridors in Montana

In another example, Georgia Tech's School of Computational Science and Engineering worked with USFS to test an approach to reconnect fragmented wildlife habitat for grizzly bears and wolverines in Montana. The goal was to create a cost-effective method to optimize corridor design for conservation planners. Mathematical modeling and open-source tools were used to analyze the trade-offs of using only the most pristine ecological lands for wildlife corridors versus combining that criterion with economic restraints. This study sought to achieve connectivity for wildlife that was also affordable. They analyzed corridors for the two species, the grizzly bear and the wolverine, both separately and combined, to determine how ecological and economic factors might affect corridor land management decisions.

In this example, the outcome that prioritized corridors for both the grizzly and wolverine species reduced the conservation costs from $31 million to $8 million. This outcome attempted to balance the economic and ecological needs of connectivity conservation, providing economies of scale for wildlife and land managers. Again, there are caveats and trade-offs for this approach, and much research exists on the topic. It's worth exploring the existing research for more information on how best to approach this type of multispecies optimization modeling.

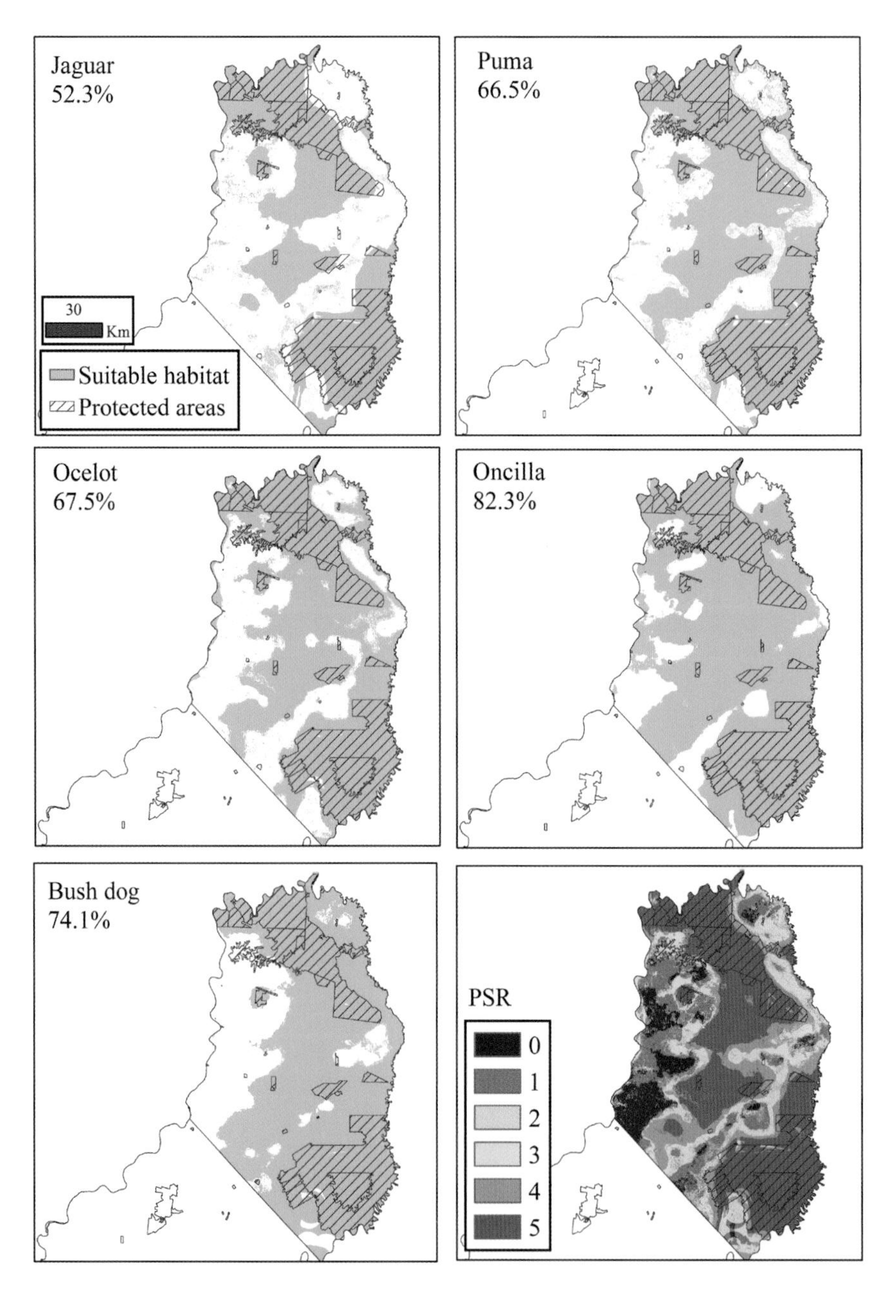

Figure 4.9. The five species-specific ecological niche models with percentage of suitable habitat across the northern-central zones. The potential species richness (PSR) map shows in dark green the areas defined as suitable for all five carnivores—jaguars, pumas, ocelots, oncillas, and bush dog.

(DeMatteo et al. 2017)

Wildlife migration in Southern California

The South Coast Wildlands led a highly collaborative interagency comprehensive planning process that produced a strategy to maintain and restore critical habitat linkages while maintaining the quality of life for residents. The plan, called *Missing Linkages: A Wildland Network for the South Coast Ecoregion*, encompasses a large area from north of Los Angeles extending 190 miles into Baja. The region covers 8 percent of the state of California and is an important movement corridor for many species. The report focuses on roads and human development as the main impediments to wildlife movement and includes detailed maps and strategies for sections within the ecoregion.

Figure 4.10 shows the Santa Monica–Sierra Madre connection areas that help decision-makers determine where certain interventions such as highway over- and underpasses and culverts can support wildlife movements and avoid vehicle/wildlife

Figure 4.10. This image shows the 15 priority linkages in Southern California. The linkages were identified through a collaborative comprehensive planning process called the South Coast Missing Linkages project. The linkages connect protected lands to create corridors that span from the ocean to the mountaintops.

SC Wildlands 2008.

Figure 4.11. The Rocky Peak Overpass, roughly 60 feet wide and 130 feet long, connects Santa Susana State Historic Park south of State Route 118 with Rocky Peak Park to the north. Mule deer, coyote, bobcat, raccoon, and skunk have been recorded using this structure.

Kristeen Penrod, SC Wildlands 2006.

incidents. Figure 4.11 shows the Rocky Peak Overpass, connecting Santa Susana State Historic Park with Rocky Peak Park.

Many of these structures already exist, and interagency partners and researchers have been tracking their effectiveness. The study offers recommendations to improve connectivity by reducing the effects of roads, rail lines, and stream, agriculture, and recreation barriers. They provide a detailed overview of their approach to building the linkage conservation plan (figure 4.12).

Modeling functional connectivity for Mexican wolves

Mexican wolves need a connected landscape to thrive. In this connectivity analysis, focused on the Greater Gila Bioregion that spans southern New Mexico and Arizona, the goal was to inform or prioritize wilderness proposals by identifying

Figure 4.12. The map shows the linkage design with the yellow outline from the Santa Monica Mountains at the coast to the Sierra Madre mountain ranges in the Los Padres National Forest. This is one of the last remaining coastal to inland connections in the region. The linkages were developed based on the habitat and movement needs of 20 focal species such as deer, badger, and mountain lion.

SC Wildlands 2008.

where there is enough potential movement for Mexican wolves to minimize genetic isolation and support seasonal migration and movement, which leads to successful recruitment of new populations.

The Bird's Eye View GIS analysis combined open-source software QGIS, SAGA, R, and Circuitscape to create the final wilderness proposal map. The resistance surface was developed by combining land use/land cover, terrain ruggedness, and distance to roads in QGIS. Circuitscape was used to analyze potential wolf movement across the landscape by using designated wilderness areas and wilderness study areas within current wolf range as the focal areas for the analysis. Linkage Mapper was used to model individual least-cost path linkages between focal areas using the

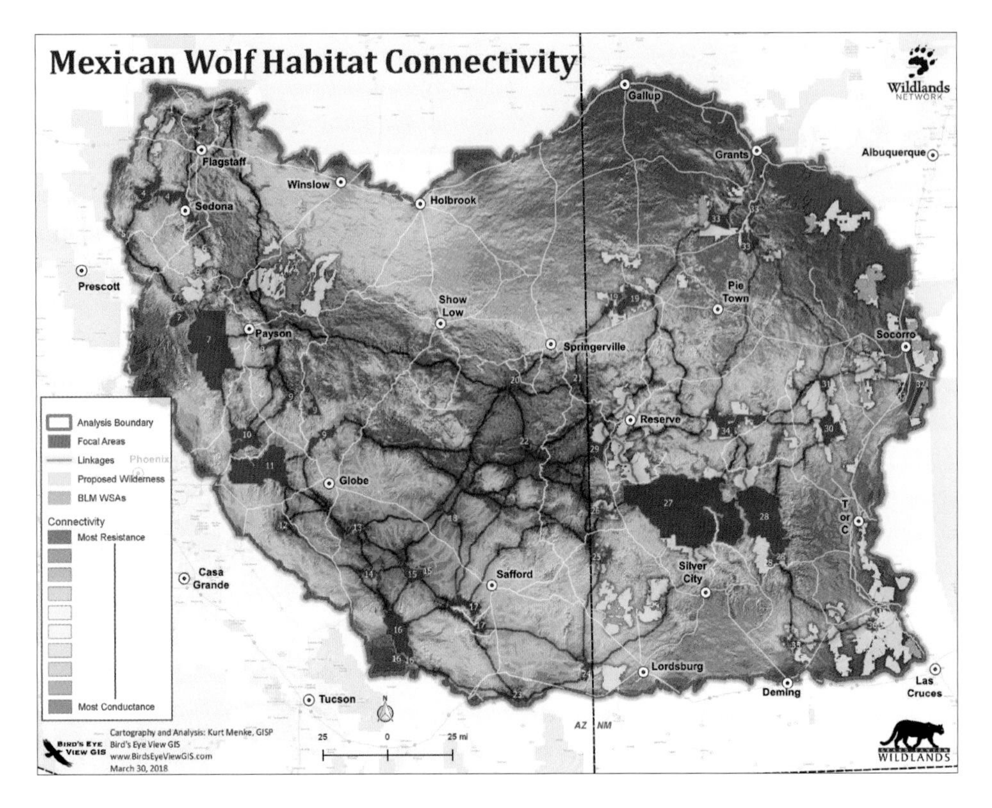

Figure 4.13. *Mexican Wolf Habitat Connectivity* map for the southwest New Mexico and southeast Arizona region.

Bird's Eye View GIS.

same resistance surface. The resulting layers included a connectivity map for the entire study area, along with least-cost path linkages that will inform the prioritization process for potential wilderness area protections in the future (figure 4.13).

Fish migration in Oregon

The US Fish and Wildlife Service (USFWS) manages wildlife refuges for corridor connectivity and is helping restore and reconnect fish habitat. The estimated 6 million dams, culverts, and straightened and ditched waterways create impassable barriers for fish such as salmon and trout that need to swim upstream to spawn. The maps in figure 4.14 show Chicken Creek, near Portland, Oregon, in its current channelized state and what it will look like once it's returned to the historic meandering path. This is an important effort on one creek that will have a big effect on the entire Tualatin River wildlife and fish corridor. The USFWS is working to open

Figure 4.14. "Before and after" maps show the now-channelized Chicken Creek (straight blue line in top image) and a visualization of the naturalized creek (wiggly blue line in bottom image) at Tualatin River Refuge near Portland, Oregon.

USFWS.

thousands of stream miles and hundreds of thousands of acres of wetland habitat to support connectivity.

Ungulate corridors in Wyoming

The Wyoming Migration Initiative uses data, science, and research to increase the public's understanding of migratory ungulates (large mammals with hooves, such as deer and elk) through outreach. The organization offers an interactive migration viewer map; cartographic products, including the Wild Migrations Atlas; and informational and inspiring videos that describe their work. One video covers the team tracking an elk herd through the entire winter migration, and another shows the effect that the 400-mile barrier of Interstate 80 in southern Wyoming has on ungulate movement (figure 4.15). The Deer 255 migration video takes you on an amazing 3D journey, following one deer's 242-mile trek and barriers navigated along the way. The satellite collar on the deer provided location coordinates that could be mapped to better understand the challenges migrating animals face and how conservation and restoration efforts are working along the Red Desert to Hoback migration corridor in Wyoming. By mapping the deer's route, land managers and researchers can better understand how it overcame obstacles such as highways and fences and how it used protected lands along its journey—demonstrating the benefits of coordinated landscape conservation efforts. The organization's work provides great examples of products that combine cartography, multimedia, science, and research to support their mission.

Figure 4.15. Elk cross Granite Creek on summer range below the Gros Ventre Mountains.

Gregory Nickerson, Wyoming Migration Initiative/University of Wyoming.

How to get started with landscape conservation planning

Esri's Green Infrastructure for the US initiative provides rich information on landscape conservation planning methodologies. The ArcGIS StoryMaps story *Seeing Green Infrastructure* provides an overview of the green infrastructure initiative. Working with a group of scientists and researchers, Esri developed data for intact habitat cores, and it is working on a methodology to develop corridors. You can access this data from ArcGIS Living Atlas of the World as well as other datasets, including ecological, demographic, cultural, and scenic data. *Green Infrastructure* (Firehock and Walker 2019) provides step-by-step instructions on how to get started with a landscape planning process, where to get the data, and how to analyze and interpret the data for a conservation plan.

Where is natural resilience the highest? And how can we identify and protect these places permanently? The Nature Conservancy's Resilient and Connected Landscapes project answers these questions. The analysis is complete for the United States except for Hawaii. As the climate changes, habitats will be affected. Some will change completely. For instance, that change might present itself as tree species not being able to withstand high temperatures or insect infestations and therefore dying off completely in areas. This will have far-reaching ramifications, as animals, birds, insects, and more rely on these trees for various resources. With rapidly changing characteristics of habitats, species will have to adapt, move, or perish.

The Nature Conservancy (TNC) found that places that have the most complexity and connectedness are the ones that will be the most resilient in these changing scenarios and where we need to protect first. Working for almost a decade with scientists and partners, TNC developed the data, science, and interactive maps to support the land protection field. The interactive resilient sites map, plus maps that tell stories for connected landscapes and conservation strategies, provide access to view and explore the data (figure 4.16). You can also download the data from the Resilient and Connected Landscapes website (www.nature.org/climateresilience) and access the detailed reports and background papers. TNC is working hand in hand with land trusts across the country to integrate this data and these tools into strategic land protection decisions. In chapter 8, see the example of how to use the freely available resilient land application to solve a real-world issue.

Figure 4.16. The TNC Resilient Land Mapping Tool showing resilient terrestrial sites in western North Carolina.

The Nature Conservancy.

All these organizations and agencies depend on partnerships with other agencies, conservation organizations, land trusts, companies, private landowners, local agencies, and others to better protect and manage corridors and achieve optimal connectivity. Land trusts can seek philanthropic and public funding to support meeting landscape and connectivity goals in partnership with others.

Join and contribute to an existing landscape collaborative

Most organizations join existing landscape efforts instead of starting new ones. Landscape efforts cover the entire globe at varying scales, led by different organizations and agencies. It's not hard to find a landscape effort to join. The International Land Conservation Network and the United Nations are good places to start to find out more about global landscape collaboratives. In the United States, start with the Network for Landscape Conservation, an umbrella network with a purpose of advancing and implementing landscape conservation at scale. The Landscape Conservation Cooperative Network (LCCN) has a wealth of information about conservation strategies, data products, planning efforts, and more. The program includes 22 autonomous regional conservation cooperatives and was sponsored by the US Department of the Interior until funding was withdrawn in April 2019. It's still worthwhile to explore the resources on the network site, especially to find out more about partnerships, collaborations, data, and conservation methodologies. To find out about landscape projects in your area, perform an internet search. Once you identify the partnerships and collaborations already active in your landscape, consider how the goals and objectives of your organization fit into the landscape conservation effort(s). Figure 4.17 shows wild turkeys using a highway underpass.

Figure 4.17. Wild turkeys using a highway underpass.

What to consider for landscape conservation goals and objectives

- What GIS data or products do you have that could contribute to the landscape effort?
- What are the gaps in data, resources, products, and conservation focus?
- What data and products exist that you can incorporate into your internal strategic geospatial planning efforts and share the results back to the landscape effort?

Use ArcGIS StoryMaps, interactive GIS applications, maps, and other products to tell the conservation impact stories of your organization and how they fit into those of the broader collaborative. The Santa Fe Conservation Trust, for example, gained understanding of its contributions to the Western Wildway landscape initiative through a simple mapping effort, described at the beginning of this chapter.

Also, if you have GIS, science, wildlife biology, or other expertise on staff, consider providing technical advice, assistance, or support to the landscape effort under way in your area. Every landscape initiative needs improved data, research, science, and expanded partnerships.

GIS methods for land protection and connectivity

Map your land protection efforts

First, you will need to create a map showing your organization's land protection efforts so you can understand how your work complements and benefits from existing landscape conservation and connectivity efforts. You can do this internally by using ArcGIS Pro, ArcGIS Online, or by working with a consultant or volunteer.

Once you have mapped your land protection projects, overlay other protected lands data. Globally, use the World Database on Protected Areas. In the United States, use the Protected Areas Database of the

United States (PAD-US) and the National Conservation Easement Database (NCED). (See chapter 8 for more on these resources.) PAD-US and NCED are available in ArcGIS Online so you can access these datasets through ArcGIS Pro or ArcGIS Online or download them directly from the host sites.

Visually inspect spatial patterns of your land protection work with other protected lands. If landscape efforts in your area have GIS layers, include those layers in your map. A good resource to find layers for your area of interest in the United States is through the LCCN website; use the science catalog to search for projects and products. If you have access to parcel data, explore and identify properties that may be good candidates for conservation within your geography and within the broader landscape.

Your land protection work may fit into a bigger landscape plan by:

- Connecting areas in a corridor for species movement
- Buffering against threats such as wildfire or sea level rise
- Conserving lands for imperiled species
- Creating places of resilience from climate change
- Protecting working lands such as farms, forestry, and ranching
- Safeguarding for water quality and quantity
- Preserving important cultural lands
- Providing access to public lands
- Linking trails

Analysis approaches to identify strategic conservation opportunities

To understand the relationships between your conservation projects and landscape conservation goals, you will need different types of GIS analysis and modeling. The following section covers some of these analyses. The Esri Press book *The ArcGIS Book: 10 Big Ideas about Applying The Science of Where* (Harder and Brown 2017) is a great resource that describes different visualization and analysis approaches. Access chapter 5, “The Power of Where: How Spatial Analysis Leads to Insight,” on learn.arcgis.com.

Simple overlay analysis

If you already have a map that prioritizes parcels or areas for current and potential land project work for your organization, overlay priority data from other sources, such as priority layers from state forest plans or TNC's climate-resilient high-quality habitat data and high-priority corridor data for species. Adding this data to your map of potential lands for protection provides deeper insights into where your current and future land protection work aligns with other priorities. Share these maps and data with the landscape collaboration and become a part of the bigger effort.

Multibenefit large landscape modeling

If there isn't a landscape effort under way in your area and you want to identify lands to protect on the basis of an array of conservation benefits, the first step is to identify the conservation themes and criteria you will model. Conservation themes should be action oriented to achieve a goal, such as "protecting water quality," "providing recreation opportunities," or "reconnecting wildlife corridors." Once you've identified the overarching themes, how will you analyze or model data to create maps that guide you to the places where protection will make the biggest impact? Try these resources that provide examples of conservation themes and associated criteria: *Strategic Conservation Planning by Ole Amundsen* (Land Trust Alliance 2011) and case studies on the Greenprint Resource Hub. Both provide conservation themes and criteria information from planning efforts by land trusts and public agencies.

Examples of themes and criteria

- Protect water quality
- Protect streams, rivers, water features, and wetlands
- Protect floodplains against development
- Protect aquifer recharge areas
- Protect headwater areas
- Protect and restore wildlife corridors

- Protect intact core habitat blocks (see Esri book *Green Infrastructure* [Firehock and Walker 2019] for guide)
- Protect riparian areas
- Protect lands with biological richness and rarity

Once you identify the themes and criteria, find the data. Chapter 4 in *Green Infrastructure* has an in-depth list of landscape-related data types and sources. Many of this data can also be accessed through ArcGIS Online and ArcGIS Living Atlas.

Where do I get data for landscape planning?

Data varies by geography, scale, political boundaries, attributes, subject, theme, accuracy, and currency. In the United States, landscape conservation cooperatives (LCCs) have identified data and science needs and developed data to fill gaps to perform analysis at scale. These cooperatives have developed data such as high-resolution land-use/land-cover, biodiversity, species, habitat, vegetation, and water resource data that either didn't exist or wasn't at a scale or quality needed for landscape planning and studies. Many of this data is available in ArcGIS Online. You can search for these products, including data, models, and science resources, and find detailed information on projects in the science catalog on the LCCN website.

Other data sources

- Esri Green Infrastructure for the US initiative data on ArcGIS Living Atlas (ArcGIS Living Atlas is a good source of broader data for planning across all scales, as is ArcGIS Online)
- US Department of the Interior federal agencies—US Geological Survey, National Park Service, Bureau of Land Management, USFWS, Bureau of Indian Affairs, US Environmental Protection Agency, and others
- US Department of Agriculture—USFS
- National Oceanic and Atmospheric Administration

- Federal Emergency Management Agency Map Service Center
- National Aeronautics and Space Administration
- NatureServe biodiversity data (available in ArcGIS Living Atlas)
- University departments focused on landscape conservation and connectivity
- State and local departments of natural resources and fish and wildlife services
- Conservation and environment organizations such as TNC, World Resources Institute, American Farmland Trust, Defenders of Wildlife, and many others

Perform a landscape multibenefit GIS analysis

Building a multibenefit model requires a GIS analyst with modeling skills. The analyst needs to know how to translate goals and criteria into discrete models that produce layers to guide conservation. Each criterion will have an associated GIS model. These models nest under each conservation theme, and the results will be combined to identify the properties that are most suitable to achieve the action for that conservation theme. The overall combined layers for each conservation theme will be combined into a composite conservation priority layer for the landscape (figure 4.18). A simple weighted overlay provides the ability to insert preference or importance of conservation themes into the process.

The six-step guide to green infrastructure StoryMaps story provides an overview of how to go through a similar process focused on habitat cores.

Approaches to connectivity in a large landscape

Modeling and analyzing connectivity require advanced analytic and modeling skills and education on connectivity and landscape mapping and modeling (figure 4.19). Within the field are many lines of thought, research, and preference for types of connectivity modeling approaches that are best suited for certain landscapes and issue types. In this section, I'll provide a high-level overview of some of the most common connectivity modeling approaches, along with resources you can access to get more in-depth information and case studies on approaches.

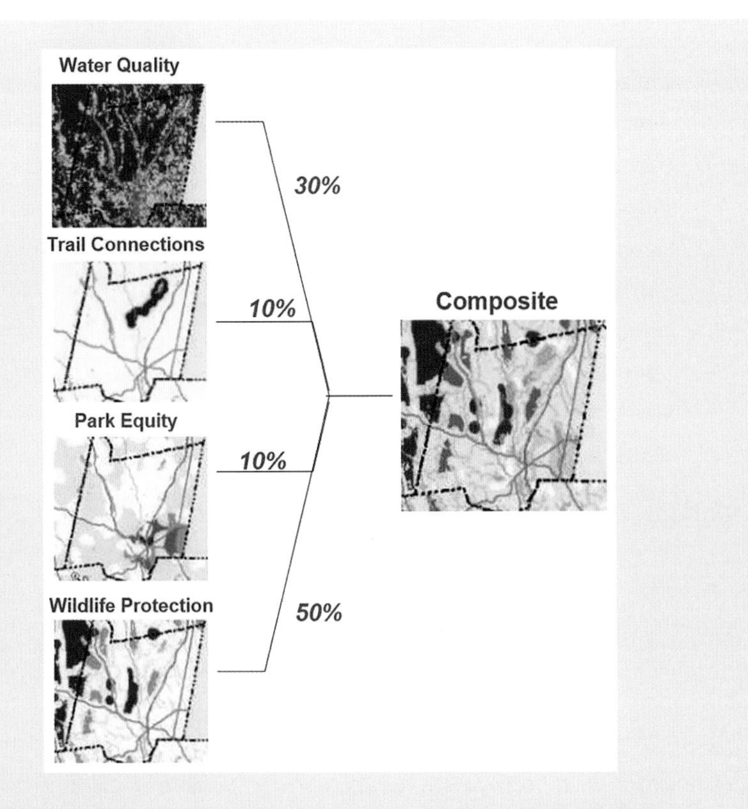

Figure 4.18. Multibenefit GIS analysis layers are combined into a composite conservation property layer for the environment.

- On learn.arcgis.com, you can access an ArcGIS Pro tutorial on how to map suitable corridors for a species. In this case, the tutorial focuses on mountain lion populations. See "Build a Model to Connect Mountain Lion Habitat."
- On LandScope, NatureServe includes a step-by-step road map to assess wildlife habitat connectivity. This 10-step process will walk the GIS analyst through process, modeling, interpreting, and implementing connectivity analysis for species.
- CorridorDesigner, created by the CorridorDesign organization, uses an ArcToolbox toolbox and ArcMap extension to provide a method for using ArcGIS to model wildlife corridors. The website includes other free tools for modeling wildlife corridors, plus peer-reviewed literature and news on the topic.

- Circuitscape is an open-source connectivity analysis software package, which borrows algorithms from electronic circuit theory "to predict patterns of movement, gene flow, and genetic differentiation among plant and animal populations in heterogeneous landscapes." This website also provides Linkage Mapper and Gnarly Landscape Utilities.
- Conservation Corridors bridges science to practice in the field of connectivity and corridors. The organization provides many resources, including up-to-date innovations in geospatial analysis data, tools, and processes.

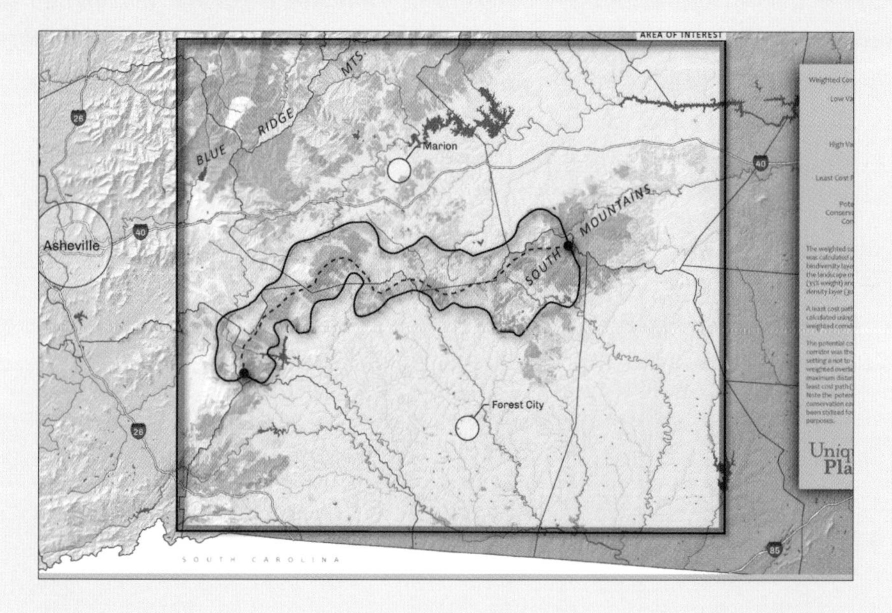

Figure 4.19. A map of a multibenefit wildlife corridor analysis in the mountains of North Carolina. The potential corridor was derived by modeling the following data: elevation, slope, hydrology, land cover, biodiversity, existing conservation, intact forests, road density, and important bird areas.

Unique Places.

Translating GIS results into landscape conservation action or recommendations

Large landscape conservation GIS analyses can result in many actions. Mapping the partnerships in a landscape can show where there is strong support and alignment and where outreach is needed to bring together more partner organizations or agencies around an initiative.

Highstead Foundation, based in Connecticut, focuses on plant and woodland conservation and stewardship. The organization plays a convening and coordination role in the Regional Conservation Partnership (RCP) Network in the northeast United States. RCPs facilitate landscape conservation across town and state boundaries in the northeast and mid-Atlantic states. The number of partnerships has grown from four in the 1990s to more than 50. The map in figure 4.20 shows the landscape boundaries of the partnerships in the region, where landscape collaborations exist, and even where they overlap. This map can be used to determine where existing partnerships could support others, where new ones might be formed, and where to focus outreach efforts to get more people involved.

For strategic landscape planning, maps can show where landscape conservation themes overlap to identify high-priority areas that need protection. The South Atlantic Conservation Blueprint prioritizes the lands and waters of the South Atlantic on the basis of current conditions using terrestrial, freshwater, marine, and cross-ecosystem indicators. Through a connectivity analysis, it also identifies corridors that link coastal and inland areas and span climate gradients. So far, more than 600 people from 180 different organizations have participated in developing the South Atlantic blueprint, and it is regularly updated on the basis of improvements to the underlying science and input from new partners. It's already helping more than 160 people from more than 70 organizations bring in new funding and inform their conservation decisions. The South Atlantic blueprint also integrates with neighboring priorities in a southeast-wide plan as part of the Southeast Conservation Adaptation Strategy (figure 4.21).

For large landscape management, maps can show where a public agency can acquire private inholdings in publicly owned lands, which streamlines management for fire regime, wildlife habitat protection, or recreation access. In figure 4.22, the

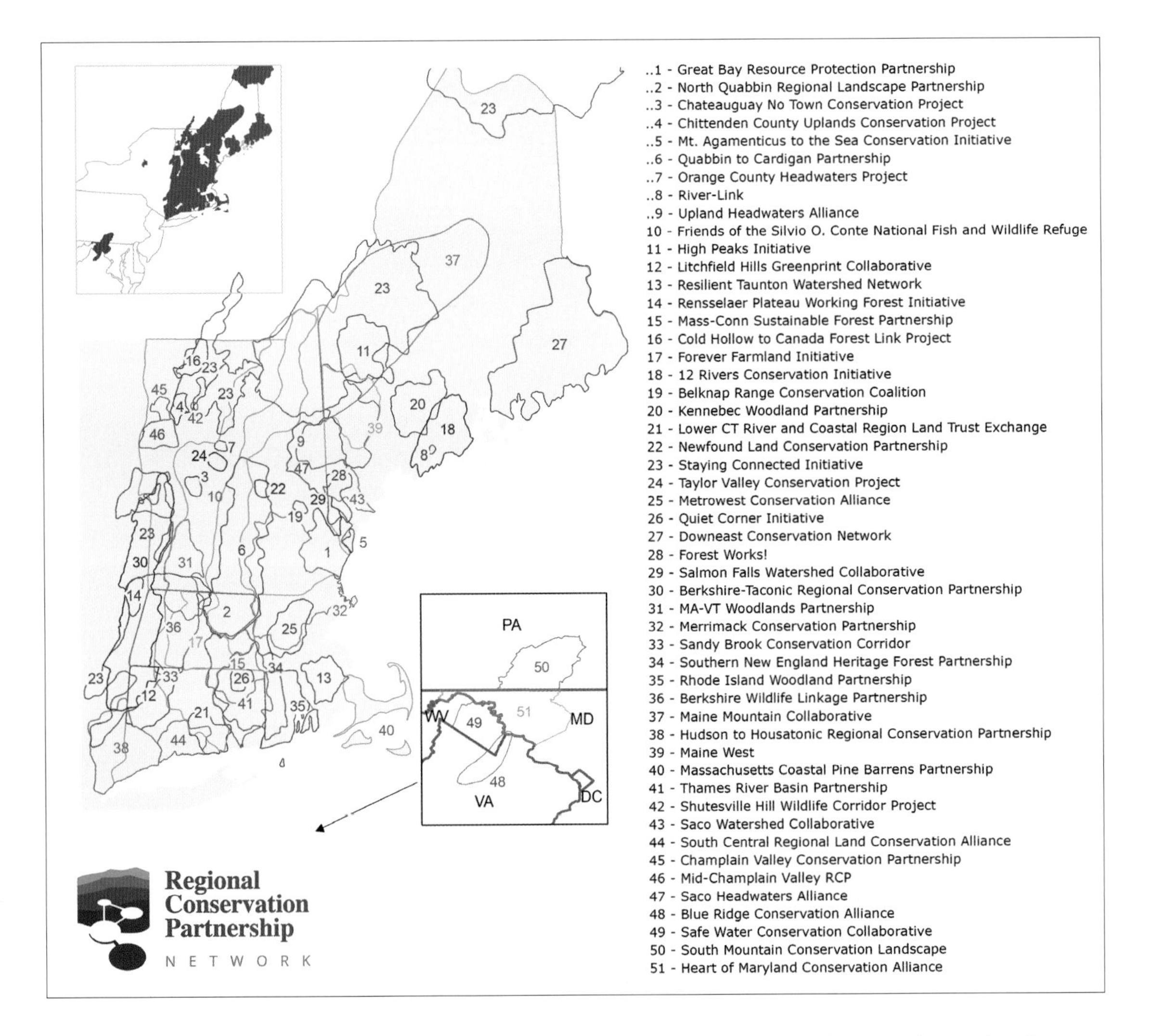

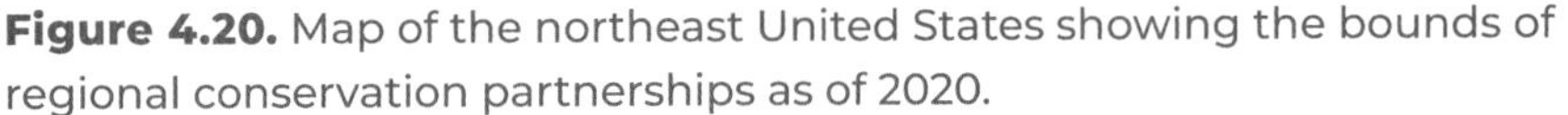

Figure 4.20. Map of the northeast United States showing the bounds of regional conservation partnerships as of 2020.

Highstead Foundation.

map shows an area southeast of Redlands, California, where you can see the checkerboard pattern of alternative public and private landownership. Land managers and conservation organizations are working to consolidate and connect these lands through acquisition or easements.

Maps can be powerful tools to educate and support policy and advocacy. For example, maps can educate policy makers on issues such as public land access for people or extractive resources on public lands and transmission lines that affect

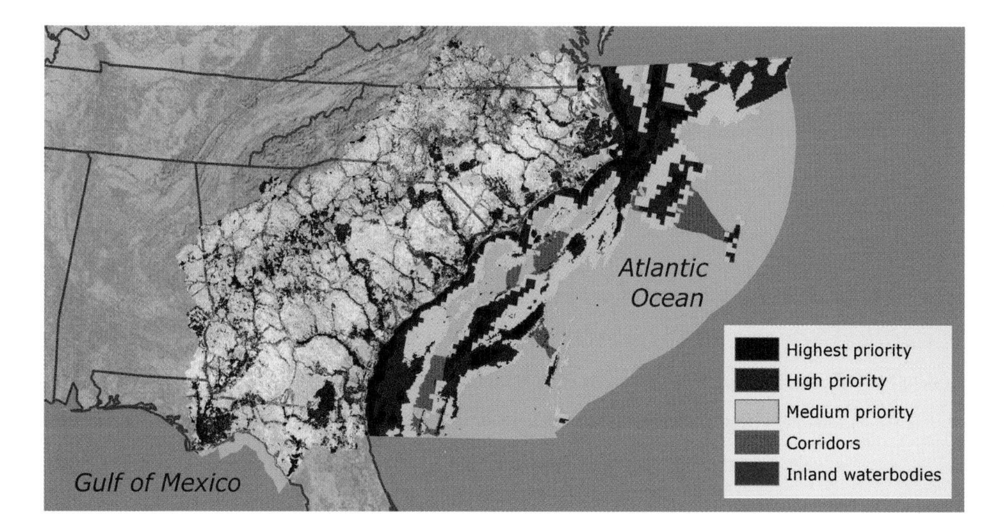

Figure 4.21. Map of South Atlantic Conservation Blueprint 2.2.

South Atlantic Landscape Conservation Cooperative. CC by 4.0.

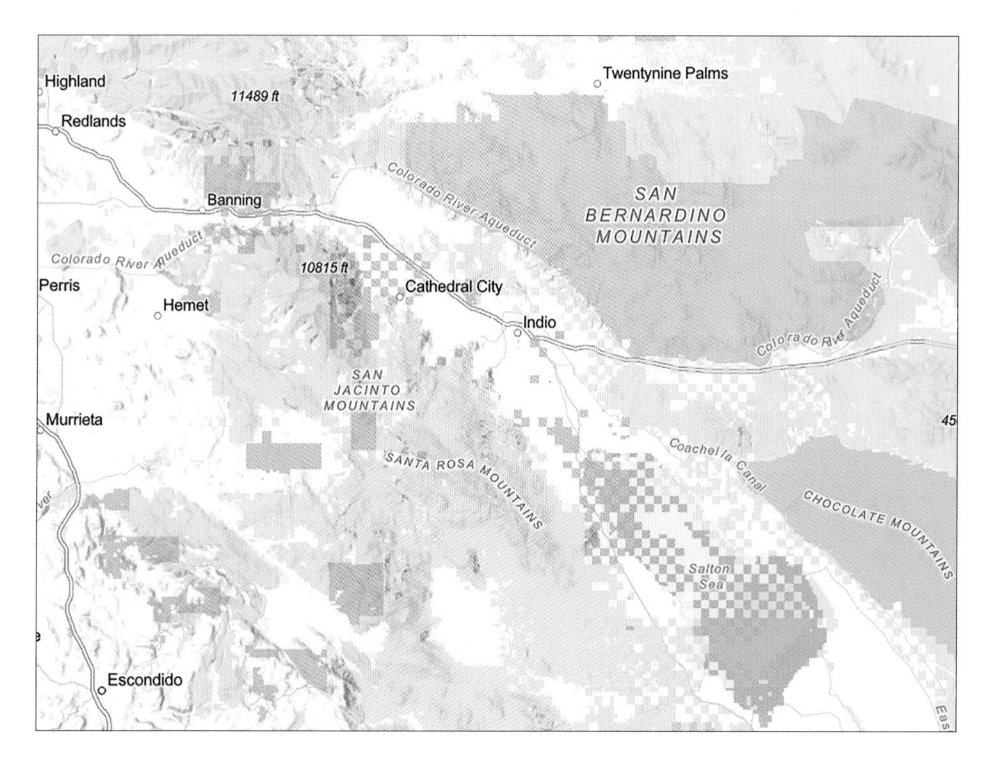

Figure 4.22. Landownership checkerboard map in the Southern California inland area southeast of Redlands.

United States Bureau of Land Management.

wildlife habitat. Maps can show where policy changes for transportation can make a big difference in wildlife/vehicle collisions by the construction of wildlife passageways under or over interstates. Maps can also make the case to change policies that affect water quality for aquatic species across a landscape.

The value of results

The results of landscape mapping efforts are particularly valuable in understanding how the impacts of local actions contribute to impacts at scale. Big targets set by efforts such as the Half-Earth Project and the United Nations Sustainable Development Goals require systematic, replicable, and defensible data, science, and mapping tools. GIS enables us to map, monitor, and measure progress at scale and across many geographic and political boundaries. Large landscape conservation and connectivity modeling provide baselines for tracking progress and supporting the coordination of partners to speed up the pace of land protection. Figure 4.23 shows mule deer using a wildlife road crossing.

Figure 4.23. Mule deer using a wildlife road crossing that includes fencing to guide the animals safely across the overpass.

© Courtesy of Colorado Department of Transportation.

How GIS improves land conservation

Typically, data lives in silos at all levels of the public and private sectors. GIS provides a common language and framework for partners across landscapes to create, combine, manage, and share data. Landscape collaborations make conservation decisions on the basis of shared maps and a collective vision for landscape conservation and connectivity. GIS provides this centralized platform in support of coordination.

Without GIS, it's virtually impossible to identify what should be protected when covering a broad geographic scale or addressing a particular concern. GIS is the organizing platform for addressing the priorities of many agencies and organizations. It provides a collaborative foundation for creating, analyzing, and sharing data to streamline conservation efforts.

GIS also provides the ability to consider opportunities and scenarios. With limited resources, land protection decisions sometimes come down to choosing one place to protect over another. GIS helps us weigh these decisions with information and scenario analysis or geodesign.

Conservationists use GIS to make the case for directing more funding and resources to the places prioritized for protection, restoration, and management. Better site selection and conservation management decisions mean greater efficiency, pace, and connecting conservation from the local to the global scale.

The Jack and Laura Dangermond Preserve

Jack and Laura Dangermond's commitment to conservation is an example of how passion and action can make a big difference. Their contributions and support for land protection organizations have ensured that millions of acres of lands are preserved and managed for conservation success (figure 4.24). Their quest is to improve the world by providing technology to support deeper insights and to make better designs and decisions. The Dangermonds signed the Giving Pledge to commit most of their wealth to three main areas, one of which is the acquisition of lands for open-space conservation and parks. Their first public honoring of this vow was in 2017 when they donated $165 million to TNC to protect 24,364 acres of one of the last remaining natural areas on the

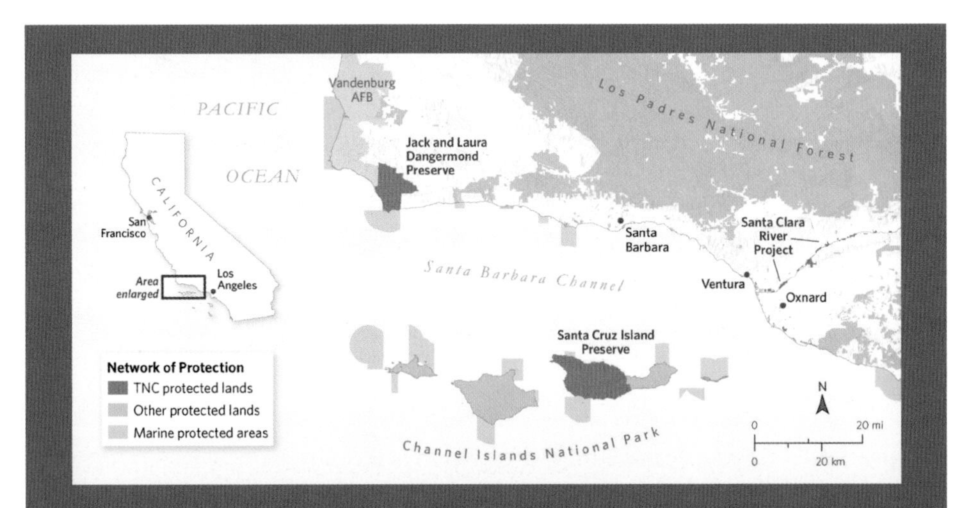

Figure 4.24. Preserve location and regional context, including proximity to protected areas.

Southern California coast (figure 4.25). This special property spans from the ocean to the mountaintops and teems with species, some of which are endangered, and habitats that provide critical benefits to nature and humans alike (figure 4.26). The land has significant cultural importance, with thousands of Chumash archaeological sites dating back more than 9,500 years. It is also rich with history, including the Spanish and Mexican era, military, maritime, and other influences.

Even though the land has been protected from development by generations of family ownership over the last century, the preserve still faces many challenges. Climate change, human development, invasive plants and animals, historical railroad construction, and oil and gas development have all negatively affected plant and animal species and, in some cases, wiped out entire populations. Still, TNC has in place a robust restoration plan that includes science, technology, education, and adaptive management strategies for the preserve. The Jack and Laura Dangermond Preserve Integrated Resources Management Plan can be found at https://www.nature.org/en-us/about-us/where-we-work/united-states/california/stories-in-california/dangermond-preserve. The plan refers to the preserve as a "living laboratory" and outlines cutting-edge and innovative conservation solutions and programs that will provide insights and information to support land protection and management globally (figure 4.27). There is also a TNC map that tells a story called *At Nature's Crossroads* that provides a rich digital experience of the preserve (figure 4.28).

Included in the plan are efforts such as launching a dune and coastal terrace restoration effort, reestablishing southern steelhead in Jalama Creek, and creating early warning detection systems for invasive species. State-of-the-art systems that include sensors and instruments to be located across the entire preserve will feed into data management

Figure 4.25. TNC's Jack and Laura Dangermond Preserve protection extends from land to sea. Its topography ranges from sea level to 1,900 feet, crossing several unique and endemic habitat types, including eight miles of shoreline, more than 300 acres of wetlands, 7,000 acres of grasslands, 6,000 acres of coast live oak woodlands, 10,000 acres of chaparral and scrub, and 78 miles of streams and riparian corridors.

Figure 4.26. Intertidal area at the preserve.

Bill Marr ©The Nature Conservancy.

Figure 4.27. Workers under giant live oaks on the preserve.

Bill Marr ©The Nature Conservancy.

Figure 4.28. Shoreline with pelicans.

©The Nature Conservancy.

Figure 4.29. Aquatic life in tide pools at the preserve.

Bill Marr ©The Nature Conservancy.

systems for modeling and management purposes. The combination of land protection with a system to restore and manage a large, special place such as this is a rare occurrence and a substantial gift to the fields of land protection, ecology, conservation biology, and related disciplines. Researchers and conservationists are deeply grateful for Jack and Laura's bold and value-driven contribution to preserve this special place as a learning laboratory (figure 4.29).

Resources

The following list of resources available to practitioners through organizations begins at the international scale and moves to the local scale.

International Union for the Conservation of Nature (IUCN) guidelines

The IUCN World Commission on Protected Areas Connectivity Conservation Specialist Group provides guidance on all aspects of connectivity conservation to support coordination at the international scale. The group's guide on best practices and connectivity recommendations, *IUCN Guidelines for Conserving Connectivity through Ecological Networks and Corridors*, is the first of its kind. It includes authoritative definitions for ecological corridors and ecological networks for conservation to guide the spatial identification of the places best suited for connectivity. It also provides guidance on planning and implementation for both practitioners and decision-makers, along with case studies from around the world. From a local to a global standpoint, this guidance will align efforts for connectivity mapping and planning so we can all scale more quickly toward a common goal and vision.

The practice of landscape conservation

The Network for Landscape Conservation is a community of practice for those contributing to landscape conservation. The network produced a report in 2018 called *Pathways Forward: Progress and Priorities in*

Landscape Conservation. This resource covers both the advances and the work yet to be done around five key areas of landscape conservation: collaborations, communications, science and data, investments, and policy. The report lays out aspirational five-year benchmark targets for the landscape conservation field to advance progress in each of these areas. The science and data benchmarks include increasing the development and accessibility of high-resolution geospatial data available to all practitioners, development of nature/culture connection data, and increasing landscape conservation applied science for climate mitigation and adaptation strategies. Figure 4.30 shows a flock of snow geese rising over a marsh in a wildlife refuge.

Figure 4.30. Snow geese lifting off in Bosque Del Apache National Wildlife Refuge, New Mexico.

Photography by Chris Peknik.

Landscape design

The USFWS report *Recommended Practices for Landscape Conservation Design Version 1.0* (2018) provides an overview of the vetted practices for landscape conservation design, the collaborative, iterative process to achieve collective landscape conservation goals. It includes case studies, linked resources, and step-by-step instructions and is a helpful guide for understanding how your organization can join broad and collaborative landscape planning efforts in your location.

Planning for connectivity

In 2012, the USFS National Forest Management Act, or 2012 Planning Rule, set forth connectivity requirements for managing forests for conservation. The planning rule calls for ecological connectivity across all land management types and is the first of its kind for US federal agencies. The requirements and approaches can be hard to understand, so Defenders of Wildlife, Yellowstone to Yukon, Center for Large Landscape Conservation, and Wildlands Network developed *Planning for Connectivity*, a guide that summarizes the ruling and approaches for anyone working on updating or contributing to forest management plans. The document provides an overview of the planning requirements, components, and best practices for connectivity planning along with case studies. This guide clarifies how to integrate the connectivity and conservation goals of your organization into forest plans.

References

Baldwin, Rob, Ryan Scherzinger, Don Lipscomb, Miranda Mockrin, and Susan Stein. 2014. "Planning for Land Use and Conservation: Assessing GIS-Based Conservation Software for Land Use Planning." Washington, DC: US Department of Agriculture.

DeMatteo, K. E., M. A. Rinas, J. P. Zurano, N. Selleski, R. G. Schneider, and C. F. Argüelles. 2017. "Using Niche Modeling and Species-Specific Cost Analyses to Determine a Multispecies Corridor in a Fragmented Landscape." *PLoS ONE* 12 (8). https://doi.org/10.1371/journal.pone.0183648.

Firehock, Karen E., and R. Andrew Walker. 2019. *Green Infrastructure: Map and Plan the Natural World with GIS*. Redlands, CA: Esri Press.

Harder, Christian, and Clint Brown, eds. 2017. *The ArcGIS Book: 10 Big Ideas about Applying The Science of Where*. Redlands, CA: Esri Press.

Meiklejohn, K., G. Tabor, and R. Ament. 2010. "Habitat Corridors and Landscape Connectivity: Clarifying the Terminology (Tech)." Accessed January 31, 2021. https://www.wildlandsnetwork.org/sites/default/files/terminology%20CLLC.pdf.

MPA News. 2001. "Paper Parks: Why They Happen, and What Can Be Done to Change Them." June 15, 2001. Accessed January 31, 2021. https://mpanews.openchannels.org/news/mpa-news/paper-parks-why-they-happen-and-what-can-be-done-change-them.

Network for Landscape Conservation. n.d. "What Is Landscape Conservation?" Accessed February 5, 2021. https://landscapeconservation.org/about/what-is-landscape-conservation.

UNEP-WCMC and IUCN. 2018. *Protected Planet Report 2018: Tracking Progress Towards Global Targets for Protected Areas.* UN Environment Programme World Conservation Monitoring Centre and the International Union for Conservation of Nature. https://livereport.protectedplanet.net/pdf/Protected_Planet_Report_2018.pdf.

UNEP-WCMC, IUCN, and NGS. 2021. *Protected Planet Live Report 2021.* UNEP-WCMC, IUCN, and the National Geodetic Survey. Cambridge, UK; Gland, Switzerland; and Washington, DC.

Wildland Ecosystem Services. n.d. "Wildland Ecosystem Services: Did You Know?" Accessed February 18, 2021. https://www.biologicaldiversity.org/campaigns/protecting_native_plants/pdfs/EcoServDidUKnowColor.pdf.

CHAPTER 5
Engaging communities to support conservation and parks

Above, The Trust for Public Land staff member Erina Keefe at a participatory design session with elementary school students at Chittick Elementary School near Boston, Massachusetts.

GIS-informed community engagement outcomes

As a conservation executive who has worked in the field for two decades, I have held fast by my number one rule for conservation and park success: engage the community. After all, we are creating parks and green spaces and protecting places to make communities great places to live. If communities aren't involved or don't support your plan, it will not succeed. Whether you are creating a plan, running a campaign to save a special place, or need volunteer support, you can find locals who are ready and willing to help. People get inspired and engaged when they see their contributions woven into the green fabric of the place they live.

I've been part of hundreds of conservation and park projects involving community engagement. I've learned that engaging the community unearths deep collective knowledge, passion, and group participation that will strengthen your work (figure 5.1). It's often the hardest part of a planning process but also the most rewarding, to see so many people get involved, or watch happy children playing in the new park that they helped design.

Figure 5.1. A "Locals Rule" sign on community sidewalk in Memphis, Tennessee.

© The Trust for Public Land.

I've also seen firsthand what happens when community members are left out. At the final meeting of a year-long community-driven conservation planning process, a prominent landowner publicly tried to dismantle and discredit the plan, compromising its credibility. We had to act fast. We met with the landowner to hear his concerns and learned he'd feared the plan would lead to the government "taking" private land. This couldn't have been further from the truth. Under the plan, places would be protected only if the landowner was willing. Once he understood and trusted the process and the maps, he came out publicly in favor of the plan, as did other community members who had been skeptical. Going forward, I learned ways to identify and address issues such as this for future planning purposes.

Barriers to community engagement

Our nation's legacy of unequal power and privilege results in mistrust that can prevent people from participating because their voices haven't been included or considered in the past. As conservationists, it's our responsibility to understand and overcome the barriers that prevent community involvement in our work. People who work multiple jobs often can't find time to come to meetings. Conservation and parks may be low on people's list of concerns. As experts, we may use jargon or get too technical, making it hard for the layperson to understand. People may not have access to transportation, feel safe or welcome at public meetings, or may not speak or read the language. We must create a multipronged strategy for community involvement (figure 5.2) to overcome these barriers, and by democratizing access to information and decision-making, GIS mapping can be a key part of this work.

Figure 5.2. Visitors to Killin Wetlands Nature Park in the Willamette Valley, Oregon, look out over the wetlands for birds such as bald eagles and wildlife such as beavers.

Metro Parks and Nature.

Participatory processes and GIS

Community engagement, when done right, builds trust, supports equity and inclusion, and strengthens cohesiveness. GIS improves public participation, and participants feel empowered and engaged, becoming cocreators and stewards themselves. Tools such as interactive visualization make it easier for both technical and non-GIS users to collaborate and contribute to conservation and park work. These efforts gain more support and pick up less friction when everyone can contribute, and GIS opens the door for more inclusion. Jack Dangermond, president of Esri, says that GIS is a "system of systems: a system of record, a system of insights, and a system of engagement"—and there are many ways GIS can be used to engage supporters for land protection.

A GIS framework for participatory approaches to community engagement

Because there are many approaches to community engagement, figuring out where to start can be overwhelming. The Community Engagement Toolkit developed in 2014 by Futurewise, Public Health of Seattle and King County, the Interim Community Development Association, OneAmerica, and El Centro de la Raza provides a detailed overview of new approaches to engage communities in local planning—especially communities that traditionally haven't engaged in planning processes because of the barriers described earlier in this section.

These engagement categories provide a good organizing framework for how GIS products are used to support participatory processes. The following section provides an overview of how GIS products are typically used during different types of community engagement.

Three categories for community engagement strategies

1. Inform and consult—get public feedback on analysis, scenarios, and results.
2. Involve and collaborate—use intensive engagement and collaboration throughout the process.
3. Empower—use advocacy training tools and fiscal processes.

GIS products for community engagement and how they're used

1. Maps—inform/consult, innovate/collaborate, empower
    - Examples—Context maps of places, maps that show prioritized areas for conservation or park creation, maps that show a particular property and why it's important, maps that show populations (human and wildlife) that are vulnerable to threats such as climate change, maps that portray a future state or design of a protected place or park (figure 5.3).
    - Uses
        - Site visits, "speakouts" (discussed later in this chapter), focus workshops, and planning events to inform the community about the current or potential state of parks and land protection. These maps are generally used to provide geographic context to community members and to show data related to the issues being addressed, such as water quality, recreation opportunity, or species protection.

Figure 5.3. A young boy participates in the community park design process to make Norwell Street greener and more climate-smart in Dorchester, Massachusetts.

        - Maps can also be consultation tools to get advice and feedback during meetings, speakouts, or other events.

2. ArcGIS StoryMaps—inform/consult, empower
    - Examples—StoryMaps stories that describe a place or an issue; stories that show conservation impacts to date and the work still to be done; stories that describe a planning process and include interactive maps of prioritized areas for protection; stories describing the people, flora, or fauna of a place; stories that describe the science behind an approach (figure 5.4).
    - Uses
        - Maps that tell a story, inform, and engage by bringing an issue to life through multimedia, such as videos and interactive maps. They can be used as a presentation platform at meetings and events or shared online.
3. Interactive maps—inform/consult, innovate/collaborate, empower
    - Examples—Interactive maps of themes such as imperiled species, water quality, wildfire threats, recreation opportunities, social equity, crime, urban heat islands (UHIs), wildlife corridors, and more.

Figure 5.4. Global Forest Watch StoryMaps story.

Global Forest Watch and World Resources Institute.

- Uses
  - Interactive maps are digital maps that allow the community to interact directly with the map, visualizing different data overlays or asking questions. Interactive maps inform the community by providing a way to explore and interpret data layers around certain issues. These maps can provide information to the community before, during, and after a plan is completed for buy-in and support. Interactive maps can be shared during events and meetings, with focus groups, and online (figure 5.5).

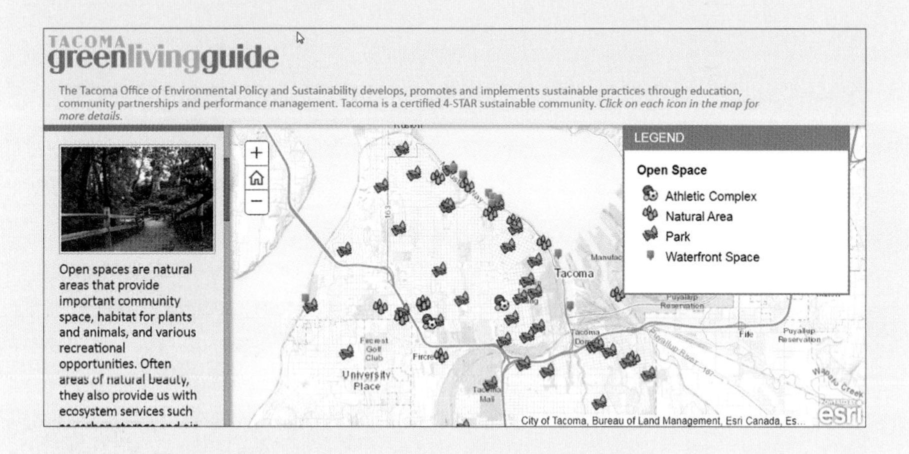

Figure 5.5. Tacoma Green Living Guide web map.

Tacoma Parks, Washington.

4. Decision support apps—inform/consult, innovate/collaborate, empower
   - Examples—Online web-mapping apps that enable data exploration, visualization, query and sorting capability, identification of features, running scenarios, geodesign (sketching features and assessing alternatives), map printing, and more (figure 5.6).

- Uses
  - Decision support apps democratize data and decision-making to empower community development of conservation and park policy and actions. These tools can be used in advocacy or other participatory process training events and shared online to reach a wider audience of supporters.

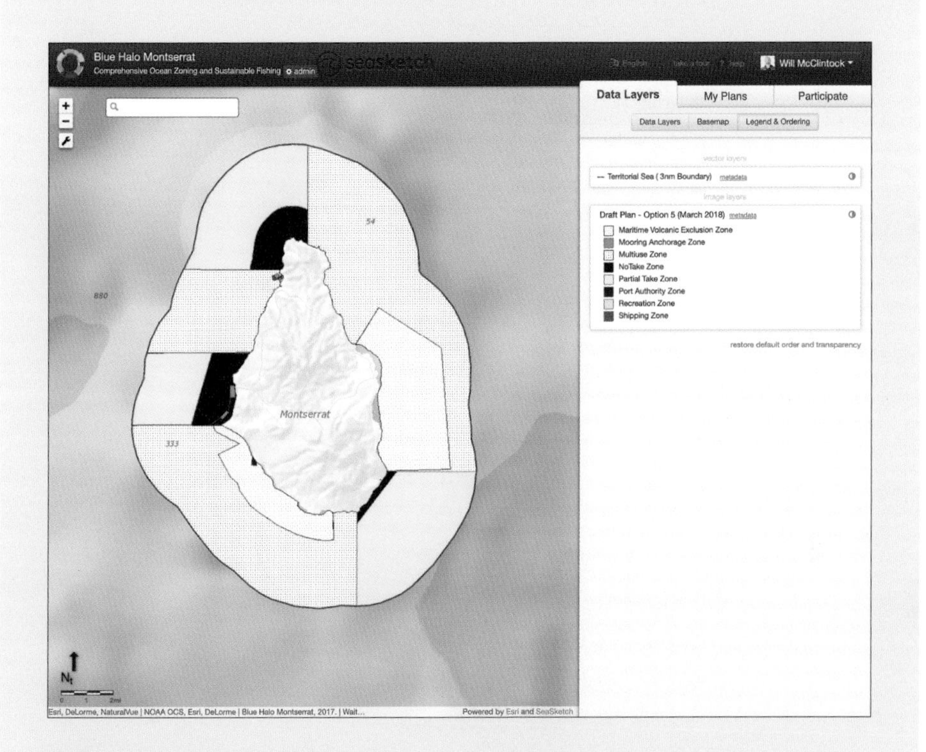

Figure 5.6. Blue Halo Barbuda—Comprehensive Ocean Zoning and Sustainable Fishing GIS decision support tool. This image shows the results of a draft plan that designates zones around the island for specific uses, such as shipping or recreation activities, and where zones would limit or ban activity for conservation or sustainability.

SeaSketch.

5. The ArcGIS® Field Maps app, combining the powerful capabilities included in ArcGIS® Survey123 and ArcGIS® Collector to explore, collect, and track information—innovate/collaborate, empower
   - Examples—Surveys that capture information on where communities want parks or protected lands and what types of amenities they want at those places, surveys on the barriers to accessing and using parks and open spaces, surveys on what the community values.
   - Uses
     - ArcGIS Field Maps can be used to capture wildlife, tree, flower, or bird data in the field; capture information about people using a park, trail, or open space; or monitor conservation easements. Targeted custom apps can focus on one topic such as sea level rise, UHIs, park finders, or birding locations.
     - Survey123 or custom apps can be used to consult with community members to provide data, information, and feedback on a variety of topics ranging from subjective input, such as, "Do you think there are enough parks in your neighborhood?" or "Where are the places you want to protect on this map?" to objective input such as, "Count the number of birds you see" (figure 5.7).
6. Open data portals—empower
   - Examples—Customizable online platforms that enable the sharing of maps, apps, data, and information about conservation and parks, as well as direct communication with users to inform them about meetings and other events.
   - Uses
     - ArcGIS Hub and open data portals inform the community about products and provide online access to data, maps, apps, and other information. The software can provide a way for the community to stay engaged and up to date on conservation and park issues (figure 5.8).

My Survey

Let's take a look at the sports amenities in this park!

Here are some top amenities. Which of the following are in this park?

- [ ] Baseball/Softball Diamonds
- [x] Basketball Hoops
- [ ] Mixed Use Field
- [ ] Paths/Trails - Paved
- [ ] Paths/Trails - Unpaved
- [ ] Skate Park
- [ ] Soccer Field
- [ ] Tennis Court
- [ ] Running Track

What is the quality of the basketball hoops?

Not Good 2 3 4 Great!

How many basketball hoops are there?

- 2 +

Figure 5.7. Survey123 parks app screen.

Figure 5.8. GeoHub home page for Douglas County, Colorado.

GIS Services Team/Douglas County, Colorado.

There is no one-size-fits-all approach to developing GIS products for participatory approaches, but some strategies and processes have a proven track record.

The Greenprint Resource Hub on the Conservation Gateway website, http://www.conservationgateway.org/ConservationPractices/PeopleConservation/greenprints/Pages/default.aspx, is an effort by The Nature Conservancy, The Conservation Fund, and The Trust for Public Land (TPL) to house resources and best practices for conservation planning processes. The Conservation Practices tab on the website offers more information on how these organizations engage diverse communities and how they use GIS products throughout each of the steps (figure 5.9).

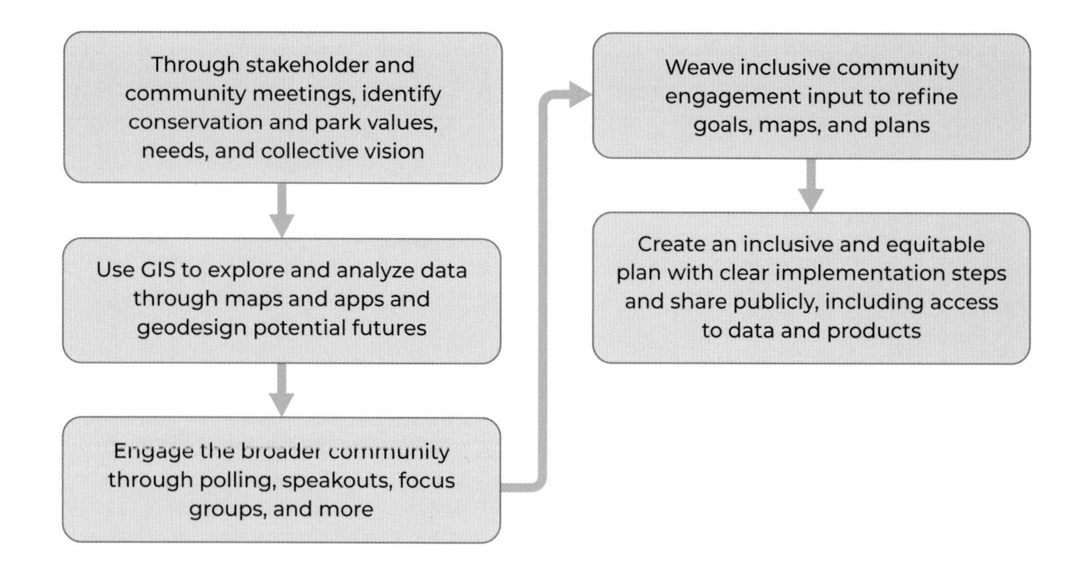

Figure 5.9. Common steps in a community engagement process.

Breece Robertson.

Putting it all together: How to create GIS products for participatory approaches

Create "working maps" for meetings or events

Most of the maps used during a community engagement process will be printed and hung on walls or put on tables. These "working maps" inform and capture information from the community. The maps themselves are engagement tools and will naturally invite interaction. They are meant to be written on, drawn on, and marked up. In chapter 1, I described the difference between internal or working maps and finished products and how professionals are best suited to make the final product maps. The good news is, you don't need a professional to make community engagement maps. You can do it.

If you want to make a map but don't have GIS capacity, where do you start? Using ArcGIS Online, you can create a free, personal account and start making maps to support community engagement efforts. For a step-by-step tutorial on making community engagement maps, go to learn.arcgis.com and search for "Getting Started with ArcGIS Online." The tutorial on creating and sharing an evacuation map in preparation for hurricane season provides the basics on creating a map in this web mapping app. You can substitute other data layers to make a custom map for your community engagement efforts. You will export the map to a PDF, and then you can print maps in the size or format you need for meetings or events.

Step-by-step guide to context maps

In half an hour, I created two simple context maps in my organization's ArcGIS Online account that could be used for a community engagement event. The first shows all parks and protected lands in the Atlanta Metro area (figure 5.10). You can try this exercise yourself by following this tutorial.

How to do it

1. Log on to ArcGIS Online and zoom in to Atlanta, Georgia.
2. Change the basemap to topographic.
3. Add content to the map. Browse data from ArcGIS Living Atlas of the World because this data is curated from trusted sources. (See chapter 1 for information on finding authoritative and curated data in ArcGIS Online.) Add the US Geological Survey (USGS) Protected Areas Database of the United States (PAD-US) manager layer from ArcGIS Living Atlas because it is the most comprehensive park and open space layer for a metro area. It includes federal, state, and local park and open space data, including TPL's ParkServe local park data.
4. Change the PAD-US layer style to a single symbol, and select green to represent the parks and open spaces.

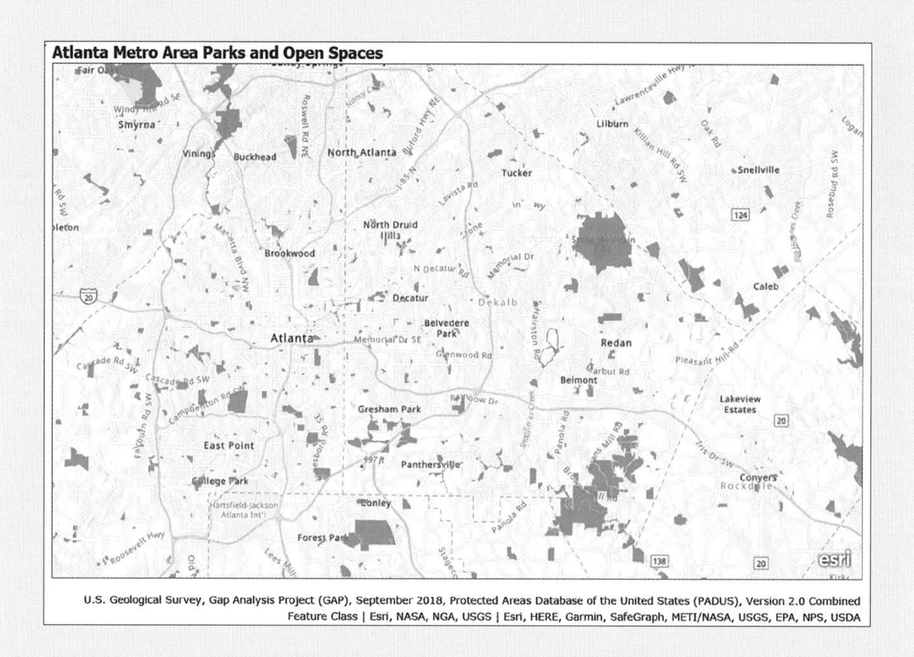

Figure 5.10. *Atlanta Metro Parks and Open Spaces* map with parks and open spaces in green.

Created by Breece Robertson in ArcGIS Online.

5. Rename the layer **Atlanta Metro Area Parks and Open Spaces**.
6. Save the map.
7. Create a bookmark so you can go back to the same map easily.
8. Click Print Map, and on the new Print tab that opens in the browser, click Save to PDF to have the PDF map printed poster size at a local print shop.

In this example, I also wanted to show where low-income neighborhoods are in relation to parks and open spaces. Going through a similar process as just noted, I searched for American Community Survey data on ArcGIS Living Atlas, found a low-income layer, and added that to the map. I styled the map to show low-income communities along with existing parks and open spaces (figure 5.11). A map such as this can be a useful tool for illustrating discussions about equitable investment in

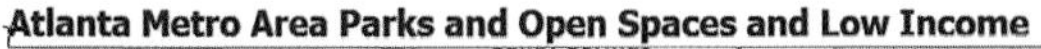

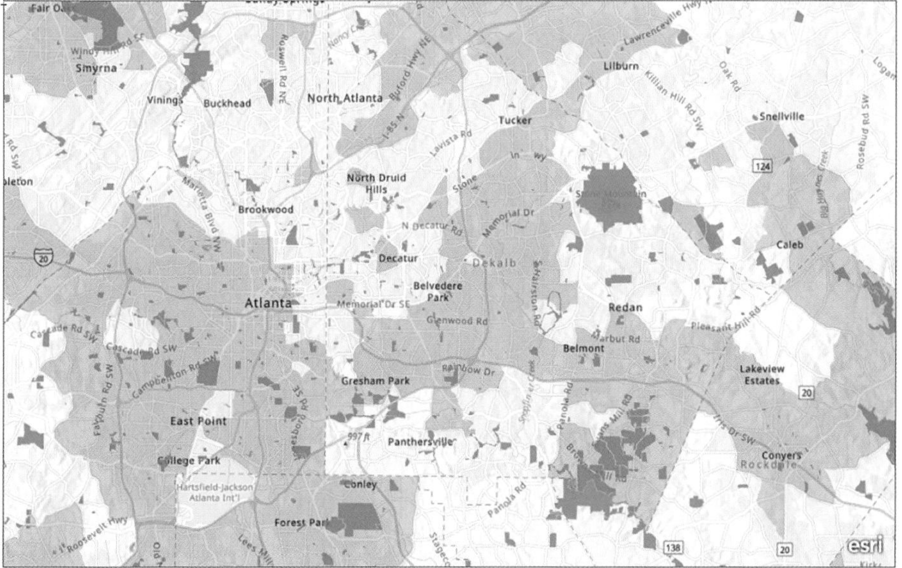

U.S. Geological Survey, Gap Analysis Project (GAP), September 2018, Protected Areas Database of the United States (PADUS), Version 2.0 Combined Feature Class | Esri, NASA, NGA, USGS | Esri, HERE, Garmin, SafeGraph, METI/NASA, USGS, EPA, NPS, USDA

Figure 5.11. Low-income areas in purple and parks in green in the Atlanta Metro area.

Created by Breece Robertson in ArcGIS Online.

parks at a community meeting. These map examples are for the entire metro area but you can zoom in and create maps for neighborhoods or specific cities. Using ArcGIS Pro, you have more options to create a map using logos, text boxes, and other map elements. You can save the map as an image file from the PDF and bring it into graphics applications to do the same. You can share the maps with others through ArcGIS Online. This is just one example of how to quickly create maps in ArcGIS Online for community engagement events or meetings.

Create an interactive map to share online

When you can't reach everyone with in-person meetings or events, you can use ArcGIS Online to create an interactive online map you can share. Here's how: Starting with the map of parks and low-income areas in Atlanta I created, I clicked Share, chose the option to create a web app, and configured a basic viewer that anyone can access online (figure 5.12). For more detailed steps on creating the ArcGIS Online maps and web maps, go to the learn.arcgis.com tutorial "Get Started with ArcGIS Online" mentioned in chapter 1.

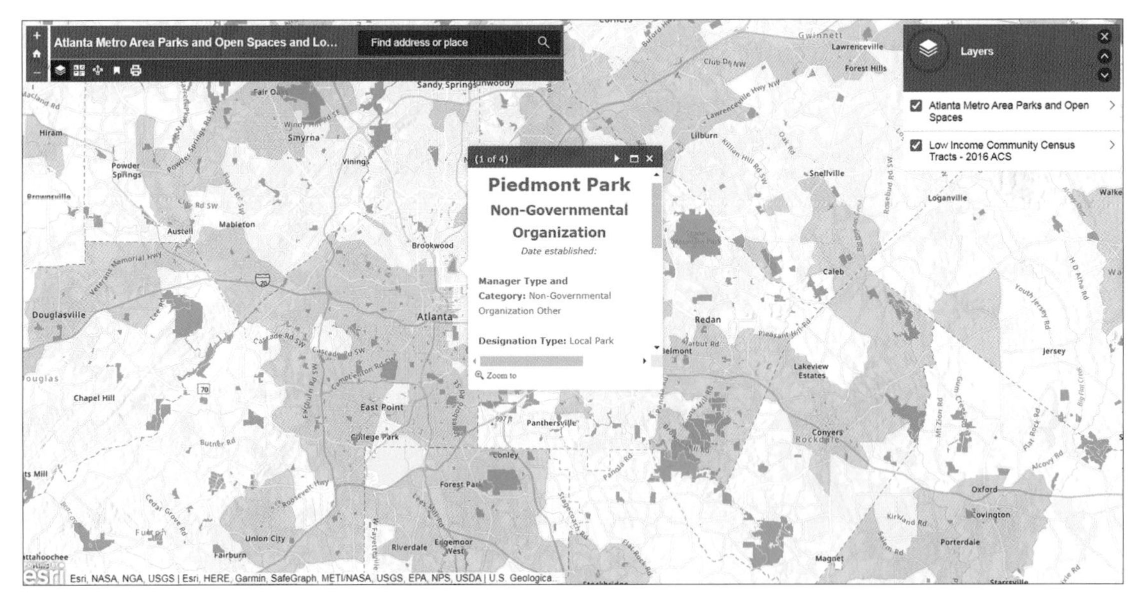

Figure 5.12. Atlanta Metro web app developed in ArcGIS Online.

Created by Breece Robertson in ArcGIS Online.

The Grand Canyon Trust tracks volunteer engagement

Interactive maps can engage volunteers. The Grand Canyon Trust tracks volunteer projects and hours and includes specific information about these projects. In this example, animals are eating young aspen trees before they have a chance to mature, which affects the habitat in the area. Volunteers put up new fencing around young aspen stands to give them a chance to grow into healthy stands to support the ecological system at Monroe Mountain, Utah (figures 5.13 and 5.14).

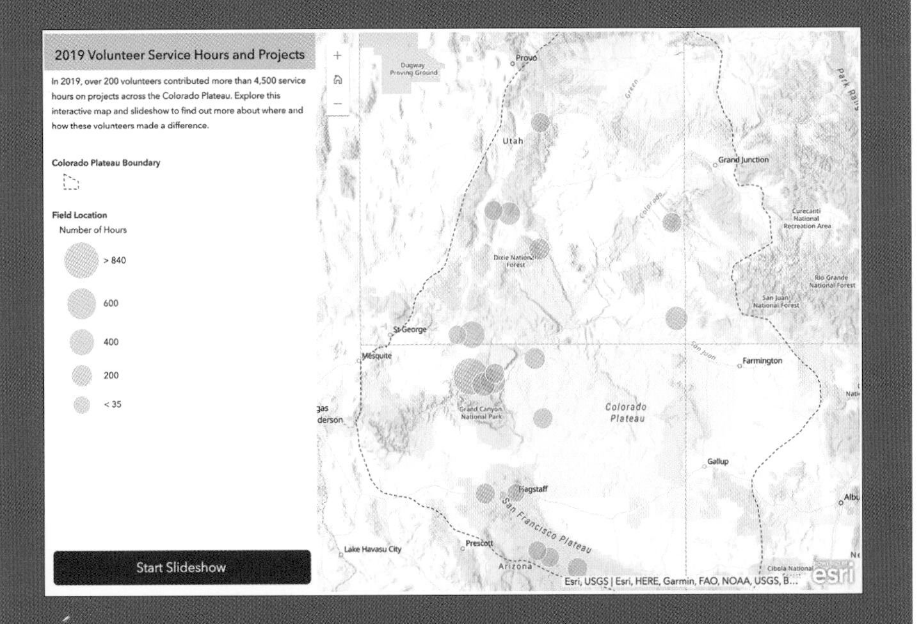

Figure 5.13. In 2019, more than 200 volunteers contributed more than 4,500 service hours across the Colorado Plateau in service of conservation projects for the Grand Canyon Trust. Each year the trust highlights these volunteer efforts and where they occurred. Using the ArcGIS Online Geo List app, custom CSS, and an interactive web map, users can explore the 2019 conservation efforts across the landscape. The initial map view gives an overview of the project location and the total number of hours of service.

Grand Canyon Trust.

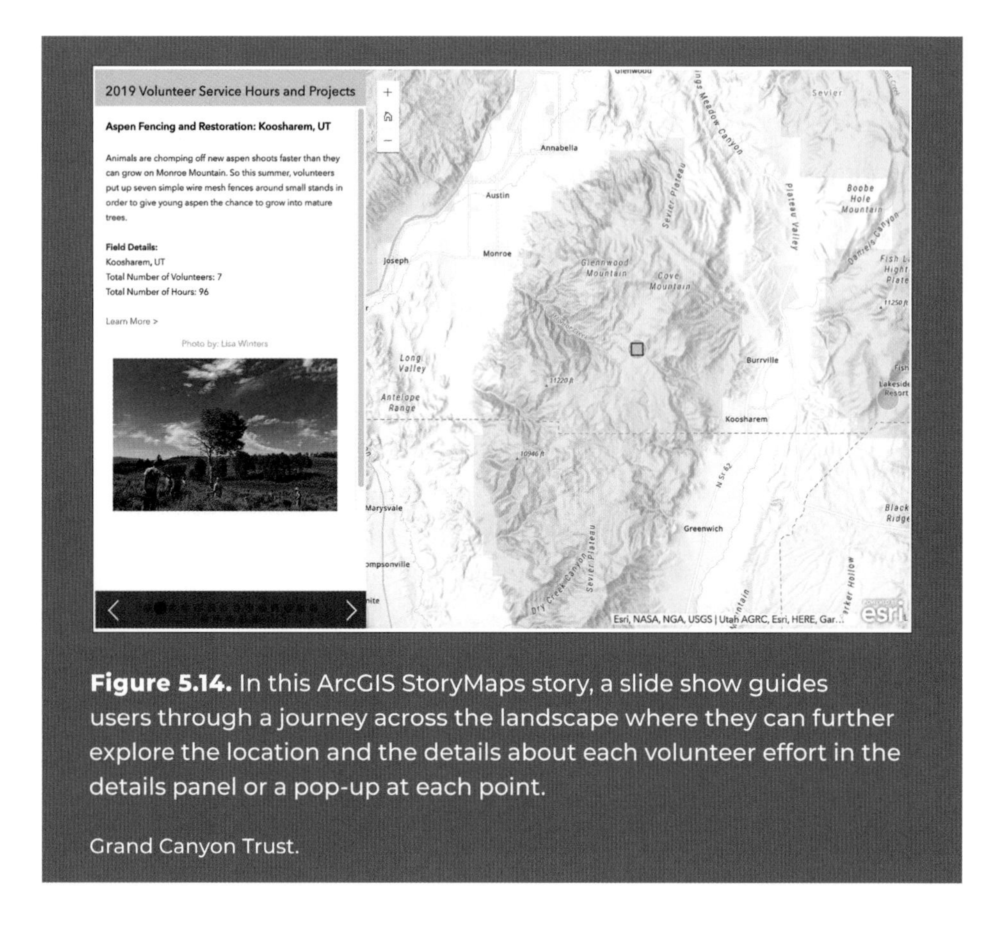

Figure 5.14. In this ArcGIS StoryMaps story, a slide show guides users through a journey across the landscape where they can further explore the location and the details about each volunteer effort in the details panel or a pop-up at each point.

Grand Canyon Trust.

Innovations in GIS and community engagement

Combining technology and engagement can require more sophisticated maps. Following are some innovative community engagement strategies that work.

Speakouts

Speakouts involve setting up a table or display at an existing event or popular gathering spot where people can learn about your work and provide feedback (figure 5.15). Bring GIS maps and question boards, and invite people to write, sketch, or use stickers and pushpins to contribute their own ideas, vision, or wishes. These maps invite people to answer questions such as the following: What do you value in your community? What places do you want to preserve? Where do you want a new park? Where do you want more trails? Where are there safety issues that prevent you from accessing a natural area? You'll need a set of high-quality maps for speakouts. This is one example of a way to provide access to information and get feedback from communities that haven't participated in conservation or park efforts in the past. And it's a great example of how GIS maps can be used in the process.

Figure 5.15. Community participants review the Taos Community Conservation Plan in Taos, New Mexico.

Real-time voting for conservation priorities

TPL provides a community-driven planning process called Greenprinting. Community members and stakeholders set conservation goals based on their priorities and the opportunities and threats in their area. TPL uses GIS to build consensus by changing model parameters and priorities and running live model scenarios during meetings. For example, stakeholders and community members who participated in the Taos Community Conservation Plan were asked to rank conservation values or themes by importance (e.g., water quality, wildlife habitat, recreation), and those rankings were integrated into a GIS model that updated the land protection priority maps. These rankings could be linked to funding sources for a particular issue, such as water quality, or associated with threats, such as habitat loss to encroaching human development. This process was powerful but it took a lot of precious meeting time to gather the information and tally it for input into the GIS models.

Keypad voting

An innovation that blended direct engagement with GIS models was the integration of keypad voting technology in GIS. During meetings, participants would get keypads and be asked to vote on conservation priorities throughout the meeting. With the press of a button, the maps would update to reflect the results from the voters (figure 5.16). This enabled TPL to get feedback quickly and accurately from large groups and incorporate that feedback instantly into conservation maps. This drastically curtailed the time spent in these meetings and focused the conversations on how the priorities of the community were reflected in the maps while encouraging everyone to contribute. This technology development helped TPL go from "old school" voting methods such as thumbs up/thumbs down or dots on maps to this new, more efficient and effective method.

Figure 5.16. In the keypad voting process, community members enter their preference or weighting for community conservation values, such as protecting water quality or wildlife habitat protection, and with the click of a few computer keys, GIS project managers can update the map for real-time discussion about conservation outcomes.

Youth and community engagement

Esri offers free ArcGIS software and products, including ArcGIS Online and ArcGIS StoryMaps, to students in grades K–12. Esri has a robust education program and many resources for supporting K–12 students in learning and using GIS. And there is a whole network of GIS professionals waiting to be mentors. These mentors guide students through using GIS to create real-world information that can help make our communities better places to live, work, and play. Students can learn how to use GIS to create maps, do analysis, and create information products to support your organization's strategic conservation work. Jack Dangermond and Will.i.am have partnered to break down the barriers of access to GIS for students and to prepare them for exciting and rewarding careers in science, technology, engineering, and mathematics (STEM) (figure 5.17). Esri also has a Young Professionals Network community to support those just starting a career in GIS or professionals just beginning to use GIS.

Figure 5.17. Jack Dangermond, *left*, and will.i.am at the 2013 Esri International User Conference (UC) in San Diego, California.

Crowdsourcing and citizen science programs for effective data gathering

Crowdsourcing is the process of collecting information or data from many people. The process can be voluntary, such as providing an app for people to collect data in the field about a specific subject, or it can tap big data sources—for example, using anonymized cell phone data to understand people's movements at parks to improve park design.

Citizen science is a subset of crowdsourcing focused on scientific research. National Geographic defines citizen science as "public participation and collaboration in scientific research to increase scientific knowledge. Through citizen science, people share and contribute to data monitoring and collection programs." Citizen science is used in the conservation and park fields to engage the community to collect data on subjects such as bird counts, wildlife sightings, and park amenities or park quality. Researchers or scientists then integrate this data into studies or data development projects that support land management, land protection, policy, advocacy, and education.

Engaging the community through these efforts can inspire new supporters for your organization or mission as members, volunteers, stewards, and donors. Apps such as ArcGIS Field Maps make it simple to configure and use out-of-the-box apps for data exploration, creation, and tracking. Community members can download these apps on mobile devices to go into the field and collect or enhance data.

ArcGIS Survey123 is a form-centric app, meaning you use forms or surveys to collect data that can be integrated into a web map or other ArcGIS apps. It also includes built-in reporting capabilities that support data analysis and display through graphs, charts, and tables.

Conservation and park organizations use Survey123 to gather information related to the conservation values of community members, park and trail use, wildlife sightings, bird counts, monitoring conservation easements, and much more. For example, on a Saturday in summer 2019 in Dallas, Texas, more than 200 community members came out to use Survey123 to collect data about park amenities and facilities to make the case for improvements to their neighborhood parks.

Some important considerations for creating a survey

- What questions will you ask?
- What data gaps will the answers fill?
- How will the information be used?

To create a survey using Survey123, you must have a publisher or administrator role in an ArcGIS organizational account for ArcGIS Online. At learn.arcgis.com, find the "Get Started with ArcGIS Survey123" tutorial. In this lesson, you'll learn how to design, develop, and share a survey and how to analyze the results, generating an online interactive map in the process.

Field Maps is a map-centric app, meaning users input data directly onto a map. This app includes the ability to capture feature data in the field, such as sketching a new protected area or identifying where new trailheads could be located. This capability requires an ArcGIS organizational account (ArcGIS Online or ArcGIS Enterprise). ArcGIS Online public accounts will allow you only to view sample or public maps. Configuring Field Maps can be involved, so you may decide to find a volunteer or professional with GIS skills to help set it up. For more information on how to set it up, see the "Resources for Configuring and Using Field Maps Capabilities" section at the end of this chapter and find the tutorial at learn.arcgis.com.

GeoForm, a form-based configurable web app, translates information to point locations on a web map in ArcGIS Online, simplifying the process of gathering information. For example, use GeoForm at meetings to capture locations people want to protect. Or have people enter an address that is displayed on a map with land protection projects to see how close they live to these places.

How land protection organizations crowdsource GIS data

Global Forest Watch, a partner-driven initiative of the World Resources Institute, is an open-source platform providing access to the most up-to-date information on forests globally. Data includes deforestation, fire alerts, and tree cover loss at different time series and spatial resolutions, much of which is near real time (figure 5.18). For more information on the Global Forest Watch dashboard app, see chapter 8.

The Forest Watcher app enables park rangers, patrollers, researchers, managers, and others to take the data and alert systems from the platform into the field on mobile devices. The app can be used offline to view and collect data.

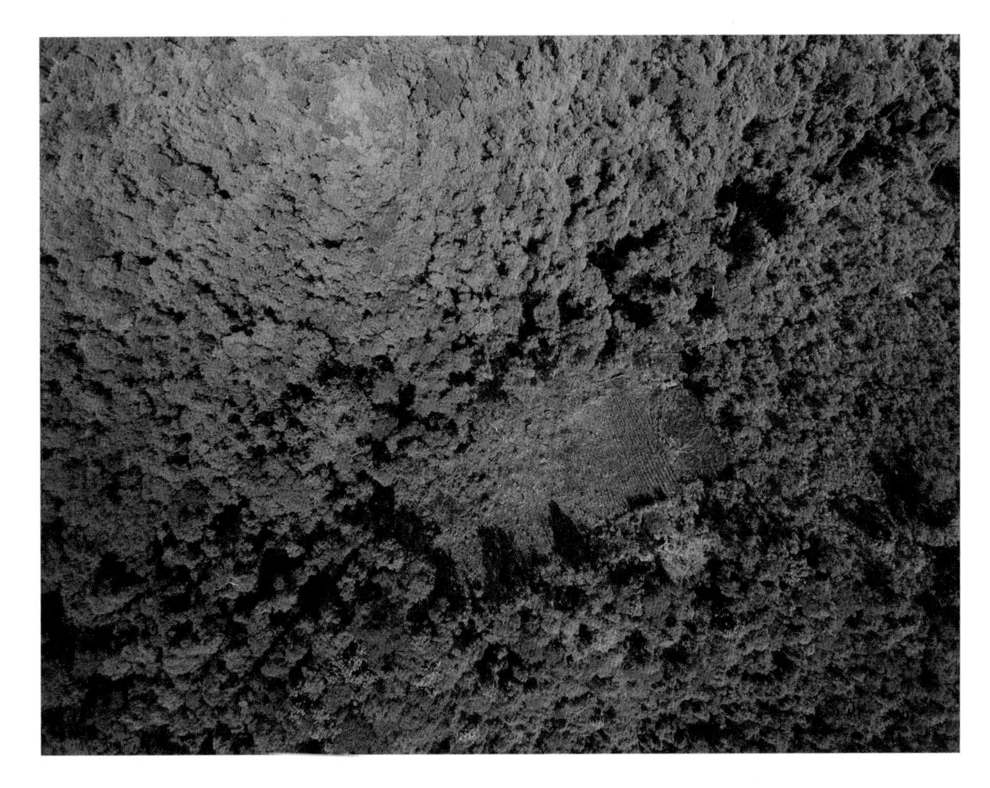

Figure 5.18. Deforestation as shown by a drone.

Global Forest Watch and World Resources Institute.

By using the app, users can investigate deforestation information they detected in the office using the mapping app, navigate to the locations, and collect information from the field, regardless of cell phone or Wi-Fi connectivity. This is an important capability, considering that many of these deforestation locations are "off grid" where there is limited or no mobile device connectivity. Users can capture photographs and GPS coordinates and record field data such as suspect causes of deforestation (figure 5.19).

Figure 5.19. Forest Watcher app in the Republic of the Congo.

Global Forest Watch and World Resources Institute.

When reconnected, they can upload the data to the mapping platform. This solution provides crowdsourcing tools linked with powerful cutting-edge real-time data for big impacts. For example, some companies that use palm oil in their products are using the data to make sure deforestation isn't linked to their supply chains. And in the Philippines, the platform was used to support the National Mangrove Forests Conservation and Rehabilitation Act, which includes restoration and protection measures for mangrove forests that provide benefits such as carbon storage, marine ecosystems, and climate buffers from hurricanes. The combination of the interactive platform and the app provides powerful tools for saving our forests (figure 5.20).

Figure 5.20. Using apps and maps in Forest Watcher in Indonesia.

HAKA.

The Sonoma Land Trust (SLT) in California uses land protection to address issues involving wildlife, climate change resiliency, and protection of habitat in Sonoma County. Land trust staff use Survey123 to collect roadkill data to complement photo-monitoring observations of wildlife moving through the landscape (figure 5.21). This geospatial information shows where animals use habitats that are crossed by roads and where land protection or management decisions can make a difference (see figure 5.22).

SLT chose Survey123 because it is easy to use, accurate, and can sync data updates in real time. One of the key reasons for adopting Survey123 was that using electronic forms on mobile devices eliminates the difficult work of transcribing handwritten paper field records into a spatial database. The app standardizes the data for incoming observations and enables SLT staff to visualize maps of the data along with graphs and summarizations. Understanding wildlife activity helps identify parcels that are essential to maintain connectivity and wildlife corridors. Revealing patterns of wildlife activity also helps land management and policy decisions. For example, if land protection isn't an option in a place with high animal/

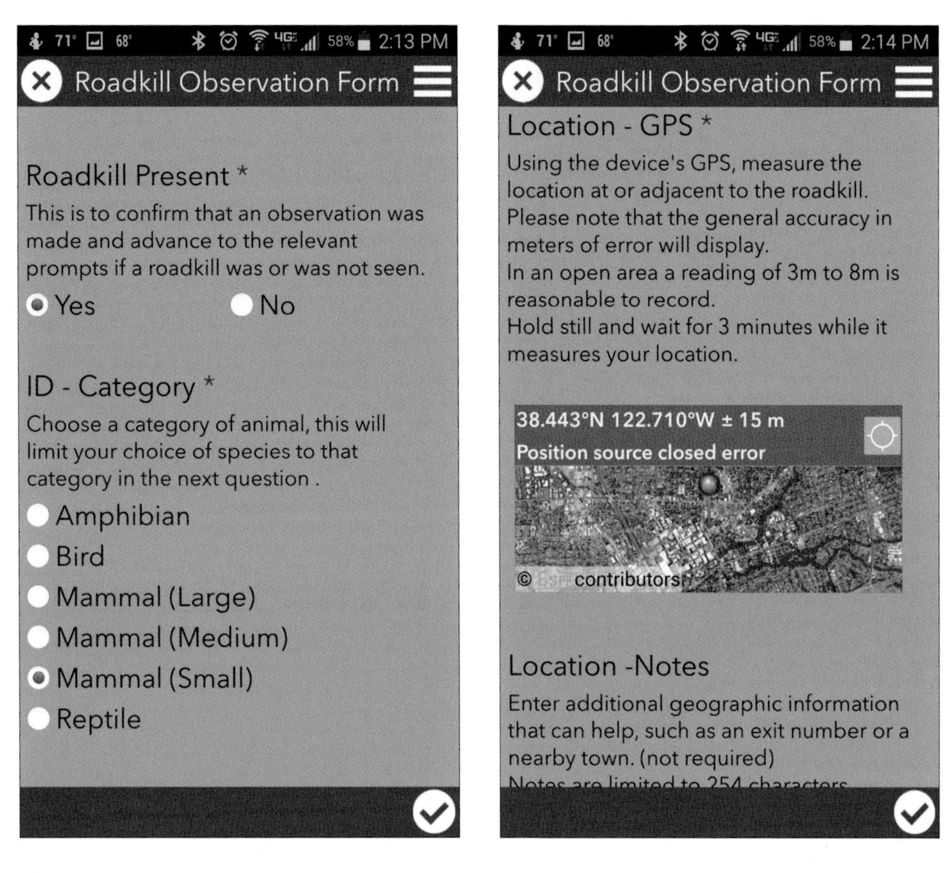

Figures 5.21 and 5.22. Sonoma Land Trust's Survey123 form tracks the type of animal killed, *left*, and requests GPS location of the animal killed, *right*.

Joseph Kinyon—Sonoma Land Trust.

vehicle incidents, the organization can collaborate with landowners to use wildlife-friendly fencing at better crossings or help transportation agencies identify locations for wildlife-friendly infrastructure such as culverts or bridges. GIS is part of the SLT land conservation toolkit to help reconnect fragmented corridors and steward the lands they own (figure 5.23).

In another example, ACRES Land Trust covers a large area across northeast Indiana, southern Michigan, and northwest Ohio. It used ArcGIS Collector to survey and manage land and habitats across 117 properties over 7,200 acres. Because of its large territory, it needed to make site visits more efficient and improve workflows in the field. It also wanted to decrease the amount of bulky equipment needed for a site visit or on-site work. Now, staff are using Collector on their smartphones

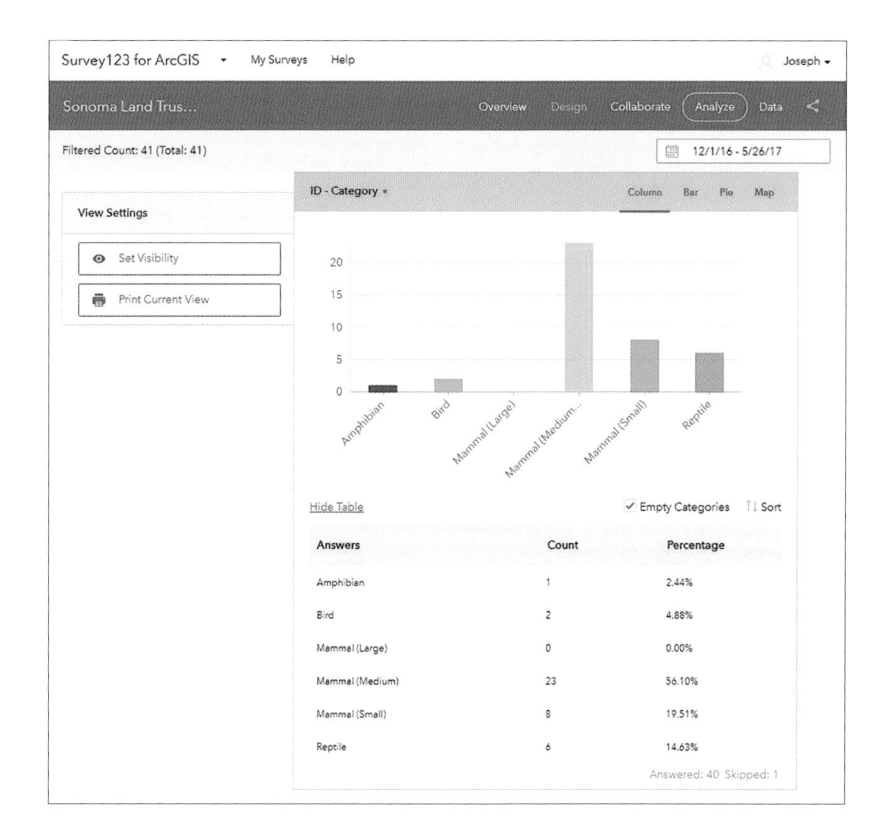

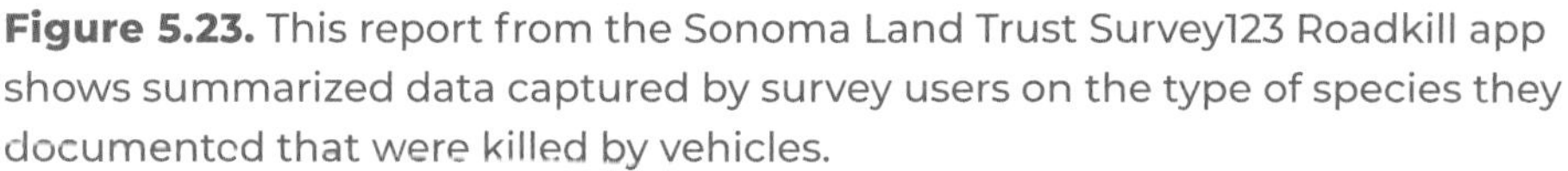

Figure 5.23. This report from the Sonoma Land Trust Survey123 Roadkill app shows summarized data captured by survey users on the type of species they documented that were killed by vehicles.

Joseph Kinyon—Sonoma Land Trust.

to capture data from site-monitoring visits on preserves, siting potential trails, and managing issues such as invasive species. The image shows a map from Collector that displays where ACRES staff, partners, and volunteers have documented and treated invasive species and allows them to monitor the effects of the treatments (figure 5.24). Whether you work for a land trust, a parks and recreation department, or a land management agency, Collector (now an integrated capability in ArcGIS Field Maps) can help you streamline data collection in the field and engage volunteers and the community.

Beyond the United States, the Snow Leopard Conservancy works with partners across seven countries to protect vulnerable leopards through landscape conservation and stewardship. The snow leopard is considered sacred in indigenous cultures and plays a big role in keeping the ecosystem healthy (figure 5.25). The snow

Figure 5.24. Collector app shows where ACRES staff, volunteers, and partners have treated invasive plants, identified by areas in pink. Data sources for map: Esri, DigitalGlobe, GeoEye, Earthstar Geographics, CNES/Airbus DS, USDA, USGS, AeroGRID, IGN, and the GIS User Community.

ACRES Land Trust.

leopard faces many threats—loss of habitat, poaching, mining, human interactions, and lack of education and awareness in local communities. The conservancy developed a Survey123 app that is used by the Land of Snow Leopard Network (LOSL) to collect data about the snow leopards that supports education, science, and research and influences policy. The app is available in different languages to encourage as many people as possible to contribute data. Some of the information captured includes snow leopard observations, prey animals, and habitat conditions.

The app also includes culture-related questions to capture perceptions from local community members on the severity and extent of threats to the snow leopard (figure 5.26). Questions encourage them to share their own ideas for how to correct these issues. LOSL is working with the Indigenous Cultural Practitioners to include Survey123 questions on culture that, when combined with observation data, can be a powerful influence on conservation policy such as the Global Snow Leopard and Ecosystem Protection Program.

Figure 5.25. The snow leopard.

Peter Bolliger.

In northern Tanzania, a key priority of African People & Wildlife (APW) is empowering community members to collect the data they need for decision-making and program evaluation. In 2019, APW began to modernize the way community members access their analyzed data using ArcGIS. The Survey123 and Collector apps (now included in ArcGIS Field Maps) are used for data collection in the field, whereas ArcGIS® Dashboards puts that critical conservation information back into the hands of community members in real time. With this knowledge, local people can visualize and use data that impacts their lives, including changes in pasture quality, thorough rangeland monitoring, and human–wildlife conflict trends and hot spots, as defined by conflict reporting. By harnessing the power of people, this type of citizen science enables discoveries that otherwise would not be possible and helps ensure the success and sustainability of our conservation programs (figure 5.27).

In 2019, APW's monitoring, evaluation, learning, and adaptation team transitioned all major data collection, analysis, and dissemination protocols to ArcGIS software. Through this new system, the team can visualize landscape-level trends and maintain a personalized relationship with each of the diverse communities they strive to support.

Survey123 for ArcGIS

My Survey

Increased

Decreased

Both Increased and Decreased

Unchanged

Other

17. How has the severity of the top threat changed in the past 5 years?

(-5 = Steep Decrease, 0 = No Change, 5 = Steep Increase)

-5 -4 -3 -2 -1 0 1 2 3 4 5

18. How has the extent (area) of the top threat changed in the past 5 years?

(-5 = Steep Decline, 0 = No Change, 5 = Steep Increase)

-5 -4 -3 -2 -1 0 1 2 3 4 5

19. Do you have any recommendations to decrease the threats you identified?

Figure 5.26. Survey123 Snow Leopard app.

Snow Leopard Conservancy.

Figure 5.27. African People & Wildlife team members in northern Tanzania using GIS on handheld devices.

African People & Wildlife/Felipe Rodriguez.

Resources for configuring and using Field Maps capabilities

Learn ArcGIS offers lessons on how to set up, configure, and use ArcGIS Field Maps. The following resources are specific to some of the capabilities in the app.

Survey123

- Learn.arcgis.com offers many lessons on how to use Survey123 to build basic or more robust surveys.
- Search for the lesson "Monitor Whales with a Multilingual Survey." In this lesson, you'll learn how to create and publish a whale survey in Spanish and English, create an app, and analyze the survey data.

Collector

- For a quick, hands-on introduction, find the "Try Collector" tutorial on the ArcGIS Blog. In this simple exercise, you will download the app and collect data about a local park.
- At learn.arcgis.com, the "Make Your First Collector map" tutorial shows how to set up the data and map used for the app in the "Try Collector" tutorial. You'll prepare a layer, create a map, and capture park assets using Collector. Other lessons at learn.arcgis.com will teach you how to use Collector when you don't have internet or cell phone service.

Why are community engagement outcomes better because of GIS?

GIS makes engagement better for the community because it improves the community's understanding of place, providing orientation and education that lead to deeper insights into problems and potential solutions. GIS maps and apps also help break down barriers to participation. For many people, these engagements are their first time seeing the place where they live or the issues at hand on a map and developing their own perceptions and ideas around why land protection is so important. It's their first introduction into thinking geospatially.

For the organization and GIS professional, maps and apps make collecting and analyzing data more efficient, whether through marking up maps by hand or creating data through an app. The GIS analyst can translate results into insights based on location, understand the frequency of events, see spatial patterns and distributions, and integrate community preferences into the map. Because the data is geospatial, you can perform analysis that isn't possible without location information. For example, you can use species data collected by volunteers using the Field Maps app and summarize whether the reported occurrences are on protected or nonprotected lands.

We've all heard the saying, "You don't know what you have until it's gone." When we engage communities with maps and apps, we all gain insights to make informed decisions to save the places that need saving and the species that rely on them (including us) before it's too late. Together, we create the world we live in, and together, we can make it better.

CHAPTER 6

Building strategy and impact into your conservation and park plans

Above, aerial view of the green hills of the Golden Gate National Recreation Area, showing the Marin Headlands, Fort Cronkhite, Rodeo Lagoon, and Gerbode Valley in Marin County, California.

What if all conservation organizations used GIS to support the creation, implementation, and evaluation of strategic conservation plans? If land trusts with plans protected twice as much land, think about how the whole conservation field could scale and benefit from data-driven strategic planning. Global visions such as the Half-Earth Project, Nature Needs Half, and the Thirty by Thirty Resolution to Save Nature, combined with local plans such as yours, could help scale the pace of land protection and its associated impacts. Similar to Half-Earth and Thirty by Thirty, Nature Needs Half is a global coalition of scientists, nonprofits, agencies, and others working to protect the most important natural places. There is a lot of momentum around using geospatial and scientific approaches to achieve these big visions.

The previous chapters covered how to map and analyze some common land protection and park themes and how to engage the community. We covered storytelling with maps and ArcGIS StoryMaps, and how to analyze data for parks, biodiversity, climate resilience, landscape conservation, and connectivity. Many of these analyses are stand-alone efforts for specific purposes. But how do you pull it all together to help drive strategy for your organization? This chapter focuses on how GIS can weave priorities for programs, initiatives, and projects together to guide strategic vision.

Why is a strategic approach to conservation important? The Land Trust Alliance (LTA) completed a census of all land trusts in 2010 and found that land trusts that have strategic plans protect more than twice as much land, on average, as those that don't. Plans focus organizations on mission-driven outcomes and help organizations lead with strategy versus reacting to opportunities. Data-driven plans help organizations understand how their work contributes to, or nests within, other efforts, such as agency programs in a watershed context, or in relation to other protected lands. Despite these important outcomes, many conservation nonprofits don't have plans. And of those that do, many don't use GIS and mapping in plan development. Written strategies and goals are important but translating these into place-based priorities supports better conservation- and park-related decisions. That's where GIS comes in (figure 6.1).

Consider the opportunity for the conservation field to scale impact by collectively creating and sharing data and plans. Of the estimated 10 million NGOs worldwide, more than 25,000 are conservation organizations. And because most are in the business of protecting, restoring, advocating for, or managing land or marine resources, the ability to map, analyze, and model geospatial data is critical to success, but relatively few do so.

In 2019, Andrew Bowman, president and CEO of the LTA, issued a call to action for the conservation field to dramatically increase land protection tenfold by 2030—the year that many scientists identify as the turning point for mitigating the effects of climate change and saving species. The Global Deal for Nature identified that 30 percent of the world's lands need to be protected to keep the average global

Figure 6.1. This map shows existing forest preserves and protected lands along with future protected areas in southeast Cook County, Illinois. This map was the result of a strategic planning process led by The Conservation Fund.

FDPCC and The Conservation Fund, Illustration by Bruce Bondy.

temperature below 1.5 degrees Celsius. To reach this ambitious goal, we must all use data and plans to collectively guide our work and track our progress.

In addition to the lands that need protection specifically for biodiversity, other lands need protection for purposes such as agriculture, working forests, urban parks, and carbon sequestration. To achieve monumental land protection goals, the LTA has a plan, including fully funding the Land and Water Conservation Fund, which was accomplished in June 2020; advocating for more conservation funding; and broadening the support base for land protection. By combining plans with steps such as these, the conservation and park field can coordinate efforts to increase the pace and scale of land protection.

But today, because of perceived or real cost or complication barriers, most organizations don't take advantage of the significant contributions GIS can make. If you're struggling to get your colleagues on board with how geospatial data and analysis can help, they might just need to see the software in motion. I have witnessed many aha moments where I showed maps or live web maps or apps that sparked insights and new approaches to problems. Some of these maps have become strategic drivers of organizational vision, initiatives, and programs.

If your goal is to track, measure, and evaluate the outcomes of your plan, GIS is the tool for this. You'll have benchmark data and maps used to develop the plan and to evaluate and track progress beyond acres and dollars. In this chapter, I'll provide a brief overview of what a strategic plan is, why it is important, and some common methods of using GIS to develop conservation plans.

Counting down to 2030

The LTA estimates that land trusts collectively conserve between one million and two million acres a year. LTA president and CEO Andrew Bowman has challenged the land trust community to increase land conservation to more than 10 million acres per year by 2030 to protect undeveloped lands while they are still available to have a substantial impact on biodiversity and climate change. Thanks to The Nature Conservancy (TNC) and NatureServe, the data exists to identify the critical lands to protect in the United States. TNC's Resilient and Connected Landscapes project (found on the Conservation Gateway website) identifies the 33 percent of lands in the Lower 48 that need to be protected for climate resilience. NatureServe's Habitat Suitability Modeling (found on the NatureServe website) identifies the 121 million acres in the Lower 48 that are most critical for biodiversity. If land trusts increase protection by 10 million acres per year, it would bring us substantially closer to the targets we need to meet collectively to save our planet. We are running out of time, and GIS has never been more important as part of the solution to meet these challenges (figure 6.2).

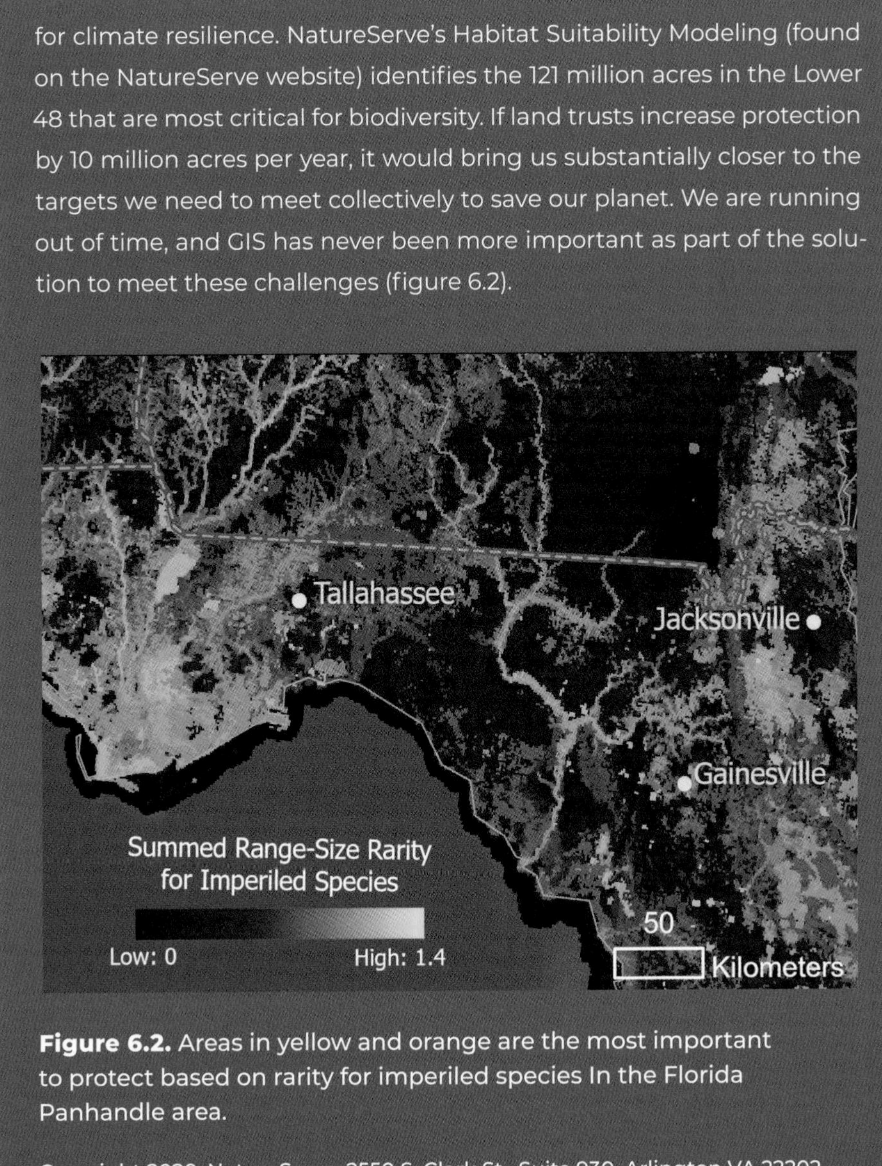

Figure 6.2. Areas in yellow and orange are the most important to protect based on rarity for imperiled species In the Florida Panhandle area.

What is a strategic conservation plan?

You can use GIS to support any type of planning, but in this chapter we'll focus on strategic conservation plans. A strategic conservation plan is a guide that establishes goals and outcomes that an organization aspires to achieve and a road map for how to work toward them, which helps get supporters and staff moving in the same direction. These plans are usually organized around the programs or initiatives of the organization.

The LTA defines strategic conservation planning as "a process that produces tools to aid decision-makers in identifying, prioritizing, pursuing, and protecting those specific tracts of land that will most effectively and efficiently achieve the land trust's mission." Ole M. Amundsen III's 2011 book, *Strategic Conservation Planning,* lays out the recommended steps for land trusts to follow when developing a strategic plan. Amundsen describes how GIS informs the steps to identify and prioritize lands to be conserved, and how GIS can also be woven into other steps for implementation strategy, financing, and partnerships.

Strategic conservation plans should be bold and visionary, but with measurable, achievable goals. For example, the strategic plan of the Land Trust for Santa Barbara County, California, outlines a goal to protect over five years 2,500 acres that provide places for recreation, wildlife habitat, and working lands. The organization also aims to protect property along the coast and potentially create a preserve. GIS will be a critical tool to identify lands that meet these specific goals.

How to pay for a strategic plan

Strategic planning takes time and money, and requires a lot of input from staff, board, and volunteers. But these investments pay off. Funders and donors want you to be strategic. They want to know you are using their money and resources as effectively and efficiently as possible to achieve your mission. Strategic plans can even be a good way to raise money for the organization. Philanthropy staff can use the plan when talking to potential donors or foundations to support fund-raising efforts.

To cover the cost of creating a plan, consider combining funding from several sources, such as foundations, agencies, and donors. A local university could likewise supply scientists, planners, facilitators, GIS support, and other key resources. This can be a learning opportunity for students and get more people engaged in your work and mission.

How strategic plans benefit conservation organizations

Internally, plans provide a way to get and stay organized and to guide and inspire staff, volunteers, and donors. By knowing where to focus conservation efforts, your organization can be strategic and proactive, directing limited resources to places with the highest conservation values. A plan enables your organization to stay clear on its mission and can also differentiate you from your peers. A good plan will set out key performance indicators and outline a strategy to track outcomes (figure 6.3). GIS provides a platform for using data to create repeatable and updatable workflows and products to track and measure conservation success.

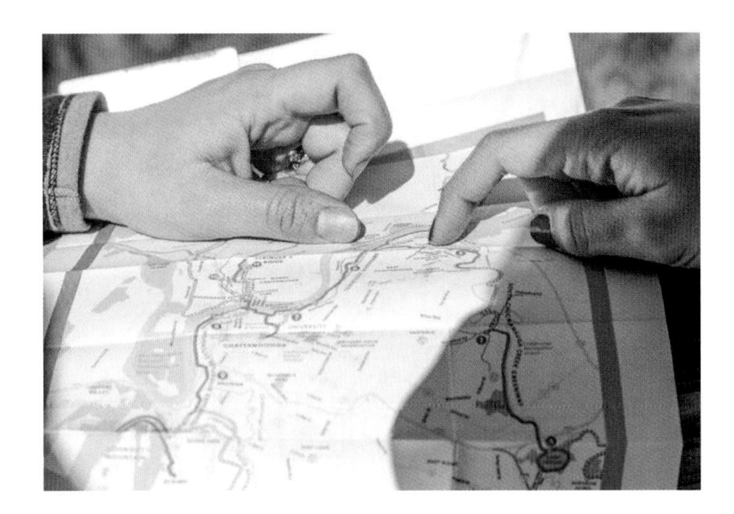

Figure 6.3. Discussing the South Chickamauga Creek Greenway map, Chattanooga, Tennessee.

Strategic plans can also help your organization say no to projects that aren't a good fit for mission or program goals. For example, if your goals are to conserve land for biodiversity, climate resilience, and recreation purposes and a property is presented that doesn't match these goals, you can back that up with data and maps and possibly refer the project to another organization or agency where it might be a better fit.

Maps and data enable your organization to respond when conservation opportunities or threats arise. You'll likely know before the issue arises because you mapped threats to land protection during your planning process and you've been monitoring the development of those threats. Or you can quickly check prioritization maps to understand how a property might fit into the plan. Maps and data can make the case for stronger protection policies or for more funding.

A plan can also help you avoid conflicts with development or permitting. Communities have used the National Conservation Easement Database (NCED) to create maps to change policy or advocate to stop pipeline development from destroying ecological and culturally sensitive lands. Some energy companies now consider conservation data in their pipeline routing or drill site planning. Good planning on the front end can save all parties time and money by avoiding land-use conflicts.

The role of easements in conservation

The NCED is the first authoritative, national database of public and private conservation easement information. I always say that if you are using only publicly protected lands in a land protection analysis, you are only including part of the story. There are hundreds of thousands of conservation easements on private lands in the nation that are key to conservation successes. Ducks Unlimited and The Trust for Public Land (TPL) collect data from land trusts and public agencies throughout the United States. This data is freely available to download from the NCED website at conservationeasement.us. NCED data is also integrated into the United States Geological Survey (USGS) Protected Areas Database of the United States product. NCED has an online, interactive web app that includes functionality such as searching for easements by easement holder or manager, identifying whether an easement is open or closed to the public, and much more (figure 6.4). NCED aids national conservation groups, local and regional land trusts, and local, state, and federal agencies in strategic land protection planning and management. See chapter 8 for an example of using the NCED online mapping app to support the protection of easements.

Figure 6.4. NCED map of the United States showing conservation easements as blue dots.

By taking a data-driven approach to strategic planning, your organization can identify where to work and why. Maps and analysis can help answer questions that are central to the mission or programs of your organization. For example: Where do people lack access to the great outdoors? Where should we focus land protection efforts first? Where are the most threatened habitats or agriculture lands? How does the amount of green space in one neighborhood compare with another, and how do these neighborhoods compare on socioeconomic characteristics such as income and race? Having this information will help guide where the organization will work on conservation in the future and how to best allocate and direct resources.

Strategic plans identify ecosystem services and the economic benefits of conservation

It's worthwhile to consider adding a conservation economics or ecosystem services study to your strategic conservation plan. Ecosystem services are the vast and varied benefits that natural ecosystems provide people, from wood to build houses to pollination of crops to clean water and so much more. A 2017 report from Charles

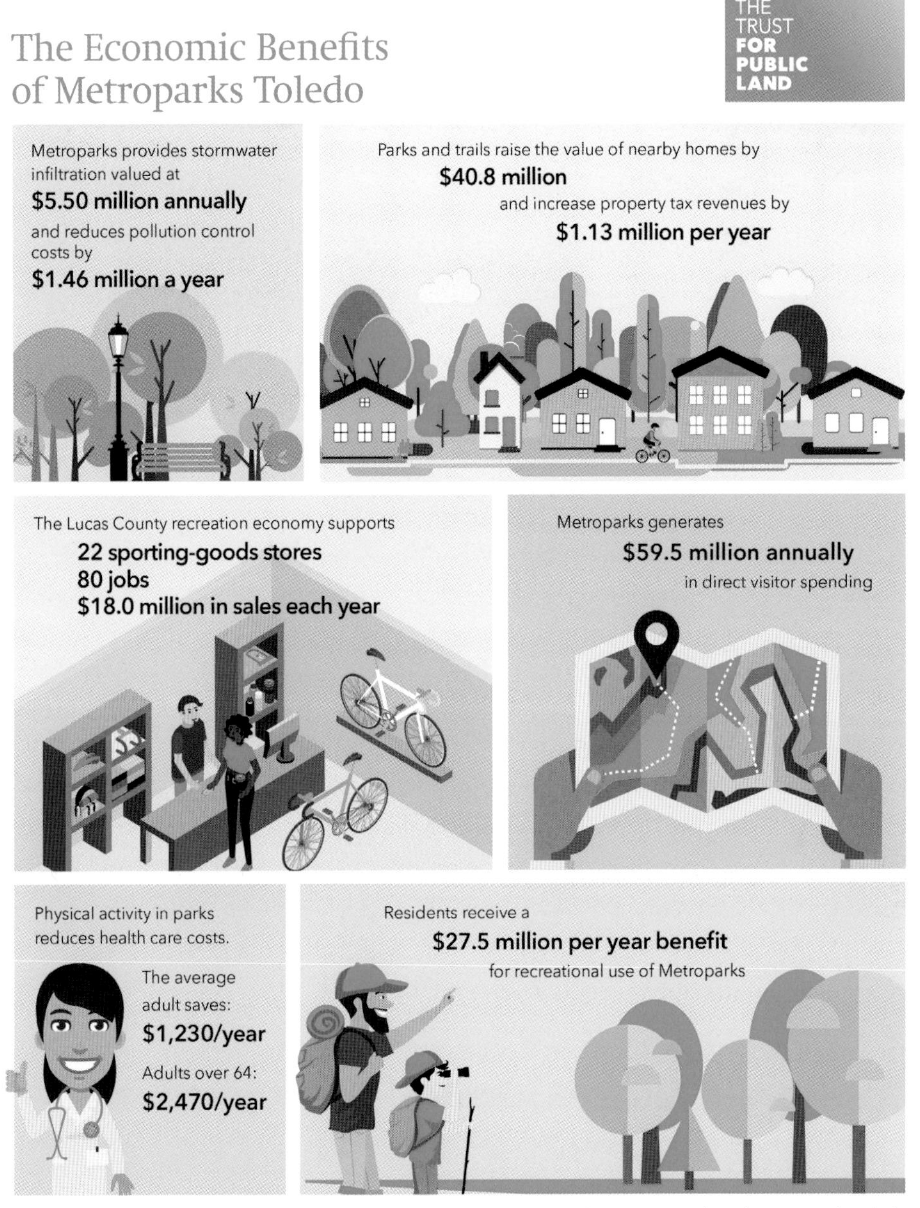

Figure 6.5. The Economic Benefits of MetroParks Toledo, Ohio, fact sheet.

County, Maryland (Department of Natural Resources), found that $577 million of the county's $4 billion annual economic activity depended on ecosystem services. A Conservation Economics report by TPL (2019) for Metroparks Toledo (Ohio) found that Metroparks generates $59.5 million annually in direct visitor spending, plus many other economic benefits, as shown in the report and infographic in figure 6.5. Conserved places, parks, green spaces, and natural systems provide significant economic value to communities, which is often overlooked as a supporting argument for conservation.

Tug Hill Tomorrow strategic plan

Tug Hill Tomorrow Land Trust covers more than 2,100 square miles in Upstate New York between the Adirondack Mountains and Lake Ontario. By 2006, the trust had protected a few thousand acres, and staff understood the landscape well and had a good sense of where they should be focusing efforts across the territory. That year, they partnered with Cornell University to develop a GIS-based strategic plan. Once the maps were complete, their strategy went from using subjective to objective decision-making for land protection. Now Tug Hill Tomorrow Executive Director Linda Garrett says GIS is part of the DNA of the land trust. They don't make any land protection decisions without using GIS as a guide. They use GIS to understand where opportunities are in relation to other protected lands, how they contribute to protecting the watershed, and how lands can help provide linkages in a landscape corridor effort (figure 6.6). They use TNC's climate resilience data to identify parcels in the "Staying Connected" corridor, a regional effort. The GIS-driven strategic plan has helped them go from a few thousand acres protected to more than 20,000 acres protected today, with many projects under way or identified.

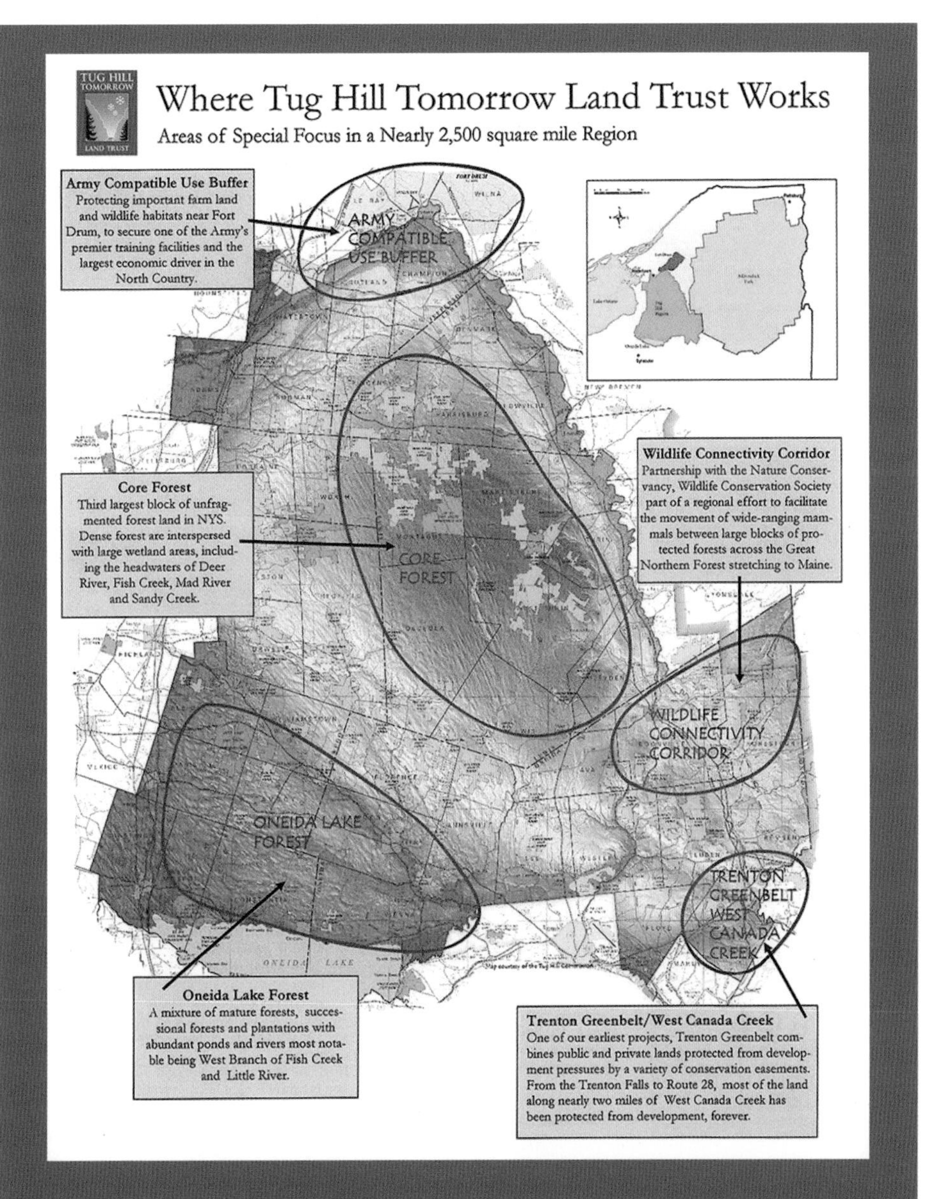

Figure 6.6. The Tug Hill Tomorrow Land Trust focal areas map includes, from top right clockwise, the Wildlife Connectivity Corridor, Trenton Greenbelt/West Canada Creek, Oneida Lake Forest, Core Forest, and Army Compatible User Buffer.

Tug Hill Tomorrow Land Trust.

Considerations for integrating GIS mapping and analysis into your strategic planning process

- Is the organization creating the plan internally or hiring from the outside to produce the plan? Examples of considerations might include whether you have internal capacity to lead a planning process or whether you can raise the money to pay for a consultant to develop your plan. Also consider whether board members or volunteers can help lead a planning process.
- How will you integrate GIS? If you have internal GIS capacity and are working with an external partner, what will you do internally versus what the partner will do? Identify the products and who owns or has access to the data and products. Get a digital copy of the GIS data and products for later use.
- If you don't have GIS capacity, will you hire a firm? How should the firm integrate geospatial approaches throughout the process based on your goals? Consider what types of products and what processes would be helpful in guiding where to work in the future.
- What conservation goals or themes will drive the plan? For example, your plan might focus on water quality, wildlife habitat, and recreation opportunities for the community. Identifying themes will help you structure the goals your organization wants to accomplish, along with the technical resources, partnerships, and funding needed to develop and implement your plan. Themes will drive the data, modeling, and analysis needed. Themes can be identified by examining consistencies in the types of lands your organization has protected, such as providing water quality benefits. Themes can be identified through a strategic planning process. Your board and volunteers can help identify themes related to needs or threats in your community. There are many ways to identify themes. It is important to identify the needs that land protection can address and match themes that will identify places that will provide the most equitable benefits and conservation outcomes.

- What types of GIS products will best support the plan? Ask yourself what kinds of maps you will need: paper, digital, online, or all of the above? And what types of stories or information the maps should show.
- Do you need a simple GIS overlay analysis or a more complex, multibenefit conservation model?
- Do you need an interactive GIS web app, such as a storytelling map or an online web mapping app?
- Will you need GIS apps that enable data collection in the field, such as ArcGIS Survey123?
- Do you need data models or methods that can be replicated or updated later?

In a strategic plan, maps, images, and data-driven infographics support the problem statement and the commitments your organization is making. Interactive apps such as ArcGIS StoryMaps, targeted location-based apps, or decision support apps make your data accessible for communities and planners alike for decision-making.

The Jane Goodall Institute: Scaling up community-driven planning and conservation using Tacare solutions and mapping tools

By Lilian Pintea, the Jane Goodall Institute

The Jane Goodall Institute (JGI) is a global conservation organization founded in 1977 that advances the vision and work of Dr. Jane Goodall in more than 30 countries around the world (figure 6.7). The institute aims to understand and protect chimpanzees, other apes, and their habitats and to inspire and empower people to become compassionate citizens and support the conservation of the natural world we all share.

Figure 6.7. Discussing progress in implementing village land-use plans and regenerating forests in Village Forest Reserves with local communities using participatory mapping and interpretation of very high resolution Maxar satellite imagery in Kigalye Village, Tanzania.

© Lilian Pintea, the Jane Goodall Institute.

The future of biodiversity is in the hands of local people. Indigenous territories hold around 80 percent of the world's remaining biodiversity, and 65 percent of the world's land is under indigenous or local community customary ownership and use. Started as a project in 1994 around Gombe National Park in Tanzania, where Dr. Goodall led her pioneering research on the wild chimpanzees, JGI's community-driven, holistic approach to conservation, called Tacare, facilitates local communities and governments to create sustainable livelihoods while planning for and advancing environmental protection.

JGI's Tacare approach addresses the threats to chimpanzee populations and habitats by assuring that local people and institutions own and drive the conservation process in their landscapes. When these local planning efforts are informed by spatially explicit and strategic conservation action plans and relevant data, there is an opportunity to connect national and global conservation goals with land-use plans that meet livelihood needs and contribute to global biodiversity.

JGI uses Open Standards for the Practice of Conservation as part of its broader Tacare approach to systematically plan, implement, and monitor its conservation strategies and actions. Conservation Standards is a science-based, collaborative, and adaptive management framework that helps focus conservation decisions and actions on clearly defined objectives and prioritized threats, and measures success in a manner that will enable adaptation and learning over time (Conservation Measures Partnership 2020).

JGI uses GIS across all steps of its Tacare and Conservation Standards action planning process.

Examples of using GIS in Tacare planning

- Defining geographic scope using participatory mapping—community mapping has two important benefits. First, it improves the quality of geospatial data to support specific conservation planning needs by recording and combining local knowledge with the latest GIS/remote sensing data. Second, it enables key stakeholders to connect early on with, and provide inputs into, the planning process.
- Mapping chimpanzee distribution—JGI uses the Survey123 mobile app to conduct chimpanzee surveys in the field and then GIS to annually model chimpanzee habitat suitability at 28.5-meter resolution covering the entire chimpanzee range in Africa (Jantz et al. 2016).
- Mapping chimpanzee habitat viability/health—defining the current and desired status or health of chimpanzee populations and habitats is a key step in the conservation planning process. This is important because it defines what is a good conservation goal and how to measure success over time (Conservation Measures Partnership 2020). In collaboration with Esri, the University of Maryland, and with support from the National

Aeronautics and Space Administration, JGI developed a decision support system by using ArcGIS apps.

- Mapping critical threats to chimpanzees—JGI has been using a variety of GIS technologies to answer the questions, "What threats and human actors are affecting chimpanzee populations and habitats?" and "Which threats pose serious problems for chimpanzees and where?" This is achieved by mapping, overlaying, analyzing, and visualizing in space and time chimpanzee populations, habitat, and viability indicators against a footprint of human activities and land-use change.
- Implementing the plan using community monitoring—in addition to assessment and planning, GIS technologies are powerful tools to help implement conservation strategies and actions and to capture monitoring data and report progress.
- Analyzing the data and adapting conservation strategies—Conservation Standards encourages practitioners to use this step to manage and regularly analyze their data as it is gathered and convert that data into useful information that could help project teams reflect and adapt their strategies and actions accordingly (Conservation Measures Partnership 2020). Figure 6.8 shows an example of using ArcGIS Dashboards to inform decision-makers annually on the status and trends in habitat indicators that they agreed to monitor as part of the Tanzania Chimpanzee Conservation Action Plan.
- Sharing lessons and stories using ArcGIS StoryMaps—the final step in the Conservation Standards cycle involves sharing lessons, giving and receiving feedback, and promoting a learning culture with project teams and with the broader community of partners and stakeholders (Conservation Measures Partnership 2020).

To enable adaptive management and learning throughout the entire conservation decision-making process, JGI is integrating its diverse geospatial data and tools as part of a one-science platform built on ArcGIS technologies. See the Tacare community-driven conservation story at https://arcg.is/DnCqr0 and Gombe 60 at https://arcg.is/H04Db.

JGI recently partnered with Esri to develop a new set of Tacare community mapping tools to enable conservation practitioners to manage and scale up their planning projects with local communities more effectively. Scaling the Tacare approach by using ArcGIS and other solutions, apps, and tools will allow for connecting, engaging, and empowering local stakeholders to better manage their lands and natural resources. In the process, it will contribute to a larger conservation vision, including protecting great apes and other biodiversity from extinction.

Figure 6.8. This image shows the Chimpanzee habitat health dashboards, which are part of the Decision Support System. The information is used to track change in chimpanzee habitat viability indicators such as habitat loss or viability to inform implementation of the Tanzania Chimpanzee Conservation Action Plan.

© Lilian Pintea, the Jane Goodall Institute.

How to approach strategic conservation planning using GIS

Understanding the answers to questions such as the following will help you scope the GIS tasks and products for the planning effort.

1. **What has your organization already protected?**

   The first step is understanding what your organization has already protected. As we covered in chapter 1, mapping the places your organization has protected might uncover new patterns, relationships, or insights.

   Next, how do these protected places fit into your organization's programs and initiatives? Using GIS data that represents conservation themes such as forests, farmland, water, wildlife, climate, and trails, you can easily see which projects fall into areas associated with a theme and which fall into multiple categories. For instance, overlaying a layer of all the public trails on the map of lands you've protected will show what you've already done to advance trail access in your community. Add a layer for waterways, and now you can see how you've benefited riparian species conservation.

2. **How do your land protection projects fit into other efforts?**

   Gather data from other organizations, agencies, and planning efforts to see how the lands your organization has protected fit into the priorities of others. You can find this data through local sources, agencies, ArcGIS Online, ArcGIS Living Atlas of the World, Data.gov, the Conservation Gateway, and many other sources. Overlay data to see the relationships of your protected places to other protected lands or priorities. The Santa Fe Conservation Trust and the Tug Hill Tomorrow Land Trust examples showed how data informed where each land trust could focus land protection efforts to provide key linkages in wildlife corridors. If you and other organizations identify places as important, it's an opportunity to partner or pool resources in an area.

3. How can you use GIS to drive organization priorities through spatial analysis?

In the conservation and park field, spatial analysis helps determine the best locations for land and marine protection and restoration, as well as analyze climate and social justice, connectivity, threats, health outcomes, and much more. Spatial analysis can also show change over time and forecast changes in areas such as land use, land cover, and sea levels. Geodesign can be integrated into a planning process to use the data created through spatial analysis in scenario analysis workflows. Scenario analysis is the process of comparing different future outcomes on the basis of knowledge and data related to current state and future influences, such as policy, funding, and potential development. It helps planners understand how to better design for future outcomes that meet the triple bottom line of economic, social, and environmental factors. Geodesign blends science with community and stakeholder values into a design process fueled by powerful scenario analysis tools that enable evaluation of design alternatives. The geodesign process supports how site selection and design of a place can best be done in harmony with people and natural systems.

ArcGIS® GeoPlanner℠ is an online geodesign tool that provides a collaborative environment to design, test, share, and iterate different scenarios for land-use planning, including land protection (figure 6.9). For more on geodesign, check the resources at arcgis.com and consider joining the annual Geodesign Summit at Esri headquarters in Redlands, California. Many universities are now offering geodesign in their curricula and offering undergraduate and graduate degrees in the subject.

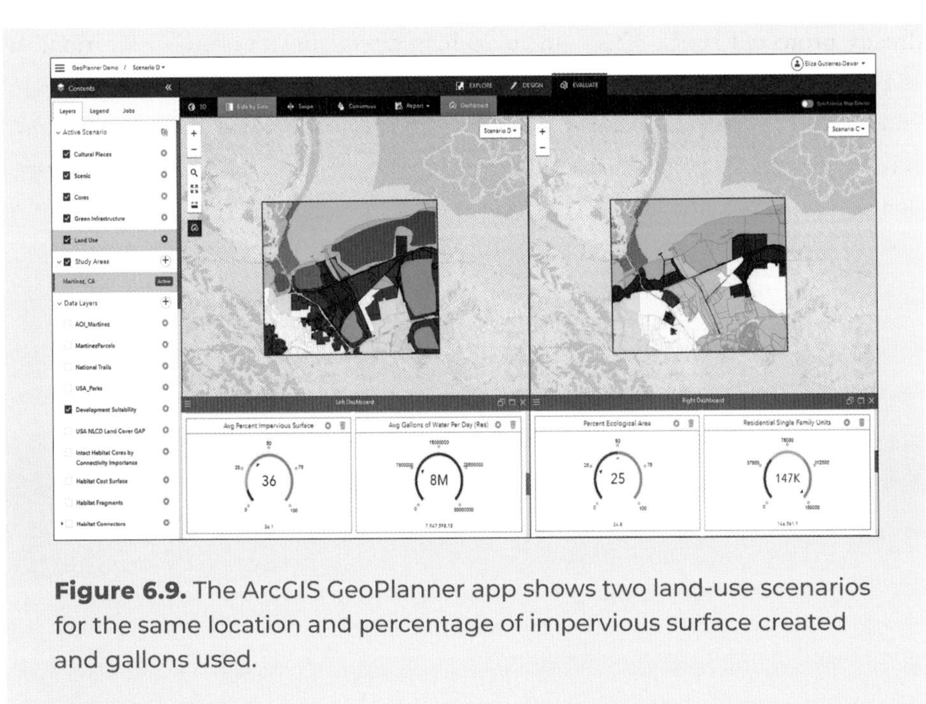

Figure 6.9. The ArcGIS GeoPlanner app shows two land-use scenarios for the same location and percentage of impervious surface created and gallons used.

The next section describes approaches that will get you started on the path to developing geospatial content for your strategic plan by using data overlays and spatial analysis.

Starting with data overlays in ArcGIS Online for simple maps

Protecting water quality is a common concern of communities and a focus of many conservation organizations. A lot of data for mapping and analyzing water quality already exists. You can simply overlay a water quality data layer with protected lands on a basemap for insights on potential actions. For example, this ArcGIS Online map in figure 6.10 displays water quality data from the Environmental Protection Agency (EPA) impaired waters data in red with protected lands in green around Mount Vernon in Washington State. This is a simple map that you could customize and style for a strategic plan to show proximity of impaired waters to already protected lands. This map could help direct conservation, restoration, or best management practices to the areas that could improve water quality. Next, you could overlay parcel data and identify vacant or publicly owned parcels adjacent to impaired water features as conservation opportunities. If a data layer exists that identifies sources of pollution, overlay this to further inform where land protection or restoration could be part of the solution.

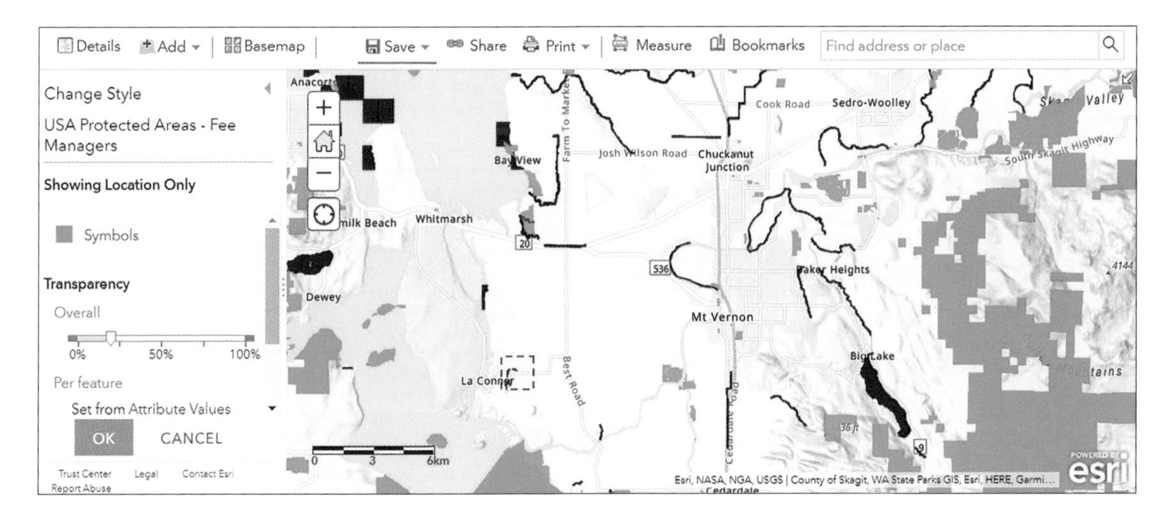

Figure 6.10. Map created in ArcGIS Online shows parks in green and EPA 303d impaired waters in red for the Mount Vernon area in Washington State. There are many areas to explore where green infrastructure parks, land protection, or restoration could help filter and clean impaired waters in this area.

Created by Breece Robertson in ArcGIS Online.

Spatial analysis

The ArcGIS Book, 10 Big Ideas about Applying The Science of Where, second edition (Harder and Brown 2017), is an important primer on understanding and using ArcGIS. The authors define spatial analysis as "a process in which you model problems geographically, derive results by computer processing, and then explore and examine those results." Geoprocessing tools in GIS apps help you manage spatial data and perform automations and spatial analysis. Spatial analysis can be as simple as applying one geoprocessing function, such as buffering protected lands to visualize potential development threats around them, or as complex as combining many geoprocessing tools into a workflow or model. The next section provides examples of how to use tools such as buffer and reclass for spatial analysis and how to combine spatial analysis approaches for suitability analyses and land protection models.

Creating buffers, reclassifying attributes, and ranking

In this section, I'll give an overview of two types of geoprocessing tools: the buffer and reclass tools. Then I'll direct you to tutorials that show you how to use these common analysis tools.

Buffers are used in spatial analysis for land protection to understand what is in proximity to already protected lands or other natural resource features such as rivers or streams. In GIS, the buffer tool creates a zone around a feature (park, stream) on the basis of distance or time. As mentioned in chapter 2, buffers can be circular or follow a network. In learn.arcgis.com, search for the lesson called "Predict Deforestation in the Amazon Rain Forest." In this lesson, you'll map the impact of roads on deforestation using analysis tools including the buffer tool.

Reclassification is used in land protection spatial analysis to group types of values in a new dataset or to simplify values on a common scale where low or high values can mean better or worse suitability for land protection. For example, reclass can be used to show where urban parks are most needed on the basis of population using a scale of 1 to 5, with 1 being "least populated" or "least suitable" to 5 being "most populated" and "highly suitable" for new parks. Of course, other information is needed to make the right decision in this case, such as how many parks might already exist in the area and whether there is a specific underserved area. Reclass is often used to create new soil and forest suitability layers. For a more detailed overview and tutorial, go to pro.arcgis.com and search in the ArcGIS Pro/Tool Reference library for "Understanding Reclassification," which provides information on what reclassification does and some common uses.

A few common examples of using the reclass tool in conservation analysis

- Reclassify land-cover data to identify wetlands, riparian areas, impervious surfaces, natural land cover, or other attributes associated with good or poor water quality.
- Reclassify stream data by stream order and combine with reclassified land-cover data to indicate where conservation opportunities are best for land types and stream order.
- Reclassify drainage attributes to natural = 5 (pervious—more absorption), soft channel = 4, hard channel = 3 (impervious—little or no absorption).
- Use the Raster Calculator tool to create a weighted sum for all input raster layers to place significance or ranking on conservation themes. For example, preserve agriculture: irrigated lands (50%), vacant lands in food deserts (25%), lands adjacent to water features (25%).

Performing complex spatial analysis: Suitability and multibenefit modeling

GIS modeling is when you combine geospatial analytical procedures such as buffering and reclassifying to gain information and insights about spatial relationships. Modeling can expose the processes and methodologies that drive the analysis and are repeatable and replicable and can also be packaged for ease of sharing. Models enable partners and the community to understand how you created a data layer to prioritize land conservation decisions. Models also make it easier for others to check and review the results. Creating models enables you to replicate and update information when new data or research becomes available. Models can be combined or run in sequence to create data products that can become inputs into other models. Modeling can be simple or complex. Figure 6.11 shows an example of a geoprocessing model that includes input data (parks, city limits), geoprocessing tools (buffer, clip), and the output dataset.

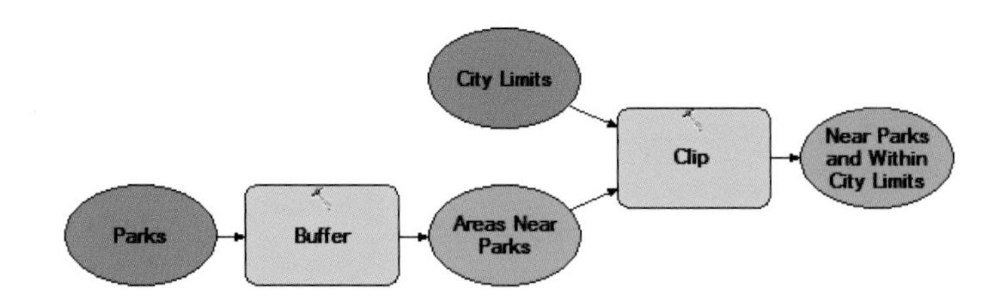

Figure 6.11. Example of a ModelBuilder™ model showing a workflow including data (in circles) and tools (in squares) to create an output dataset to identify areas that are near parks within the city limits.

Modeling approaches for conservation themes

Following are lessons on modeling approaches for specific conservation themes. For a strategic plan, you'll likely pull the outputs from thematic models into one overall model that shows the places to protect based on all the themes. An overall model result shows where multibenefit conservation can occur to address more than one conservation theme—for example, protecting water quality and farmland at the same time.

- Search learn.arcgis.com for the lesson called "Model Water Quality Using Interpolation." In this lesson, you'll learn how to model water quality in the Chesapeake Bay by analyzing dissolved oxygen levels.
- Search learn.arcgis.com for "Build a Model to Connect Mountain Lion Habitat." In this lesson, you'll learn how to build a model that identifies the best locations where roads intersect mountain lion movement corridors to build overpasses for safe passage for the mountain lions.

Common sources for water quality data

- The USGS has many data download products and interactive online maps through the National Water-Quality Assessment, covering topics from stream flows to pesticides.
- The Water Quality Portal provides data downloads from states, tribes, and local governments. This is a cooperative

effort by the USGS, the EPA, and the National Water Quality Monitoring Council.

- The EPA's Office of Water Programs provides water quality data, including 303(d) impaired waters.
- State surface water quality bureaus.
- Regional and local water quality departments.
- ArcGIS Online and ArcGIS Living Atlas serve up authoritative water quality data from federal to local agencies and organizations.

Analyzing threats

Once you've created maps to guide land protection on the basis of themes, consider including threats, such as population growth, development pressure, energy development, and climate change. Identify the threats in your study area and find those layers to add to the analysis. These threat layers can highlight what needs to be protected first before the opportunity may be lost to preserve those places. Just as with the conservation theme data, you can find that information in ArcGIS Online, ArcGIS Living Atlas, or through conservation organizations or agencies.

It's likely that a multibenefit GIS analysis will return a lot of parcels that need protection. You can further filter the parcel data by adding parcel-specific selection criteria, such as size of parcel, adjacency to other protected lands, cost, and other criteria that conservation staff rely on when making decisions about purchasing properties.

Your strategic land protection map will inform opportunities, but your staff doing the land deals will need to apply another level of filters on the basis of their knowledge of the community, on how they approach willing-seller landowner analysis, and on the right conservation acquisition tools to apply to certain properties. Maps such as these will be a key tool that conservation staff will use with the landowner in the discussion about conserving their property. Once the land is protected, organizations and agencies then must steward, monitor, and manage it. GIS is a crucial tool for selecting the right locations to protect, so you can be confident your time, money, and resources will make the most difference.

In the next chapter, we'll investigate how to track and evaluate the outcomes of land protection and strategic plans.

Strategic planning resources

- On NatureServe's LandScope website, search for "How to Prioritize Sites" for a step-by-step guide for practitioners to identify priority places for protection. This detailed guide can be applied to any conservation or park planning process. The guide covers assembling the planning team, developing the goals, obtaining the data, details on the modeling approaches, and creating the plan.
- Esri's Green Infrastructure framework is a detailed guide to approaching green infrastructure planning using a broad array of conservation themes but driven by the concept of intact habitat cores. It can be found online. There is also a book called *Green Infrastructure* (Firehock and Walker 2019) that is a great companion to the digital resources and includes step-by-step approaches for conservation and park planning.
- The LTA has a rich library of best practices and examples for how land trusts have developed and implemented plans. These resources can be found on the LTA website in its Learning Center.
- TNC, TPL, and The Conservation Fund developed the Greenprint Resource Hub as a one-stop shop that details how the three organizations approach data-driven strategic conservation planning.

How advances in GIS technology supported a community-driven conservation planning framework

Strategic plans are crucial tools for your organization's success, and likewise, communities and local governments can benefit from conservation plans of their own. TPL helps communities create "greenprints" for open space and resource protection planning. This interactive, community-based modeling process uses GIS to identify priorities for planning and conserving parks and natural resources on the basis of local input.

The process combines the science behind habitat, water quality, and other resource protection goals with the art of community engagement to identify shared values around conservation, trails, and park creation. Participatory design, including crowdsourcing data, electronic voting tools, and other community engagement approaches, such as speak-outs and polls, provides more opportunities to involve as many people as possible. When you involve the community in your planning process, you provide an opportunity to create a shared conservation vision that the community supports (figure 6.12).

Figure 6.12. A visitor adds design ideas to park photos in Rodney Cook Sr. Park, a new 16-acre park in the Vine City neighborhood of Atlanta, Georgia. It is a rebuild of a 19th-century Olmsted-designed park. Extensive public participation went into the new design, which features museums, a library including volumes from the C. T. Vivian and Martin Luther King Jr. families, youth and sports activities, and farming, among other amenities.

What greenprints provide

- A strategic data-driven conservation model
- A community and stakeholder-driven process
- Objective data to direct conservation funding and policy
- The ability to quantify trade-offs

TPL developed its first greenprint for King County, Washington, in 2004, and today, the data and methodologies still influence the way staff and partners think about building strategic conservation plans, crafting initiatives, and setting priorities. That is a long track record for the influence of a GIS-based planning process, especially when plans often tend to sit on the shelf. Online greenprint decision support applications provide data-driven, supported, and updatable tools, accessible by all, to maximize strategic, community-supported land protection efforts (figure 6.13).

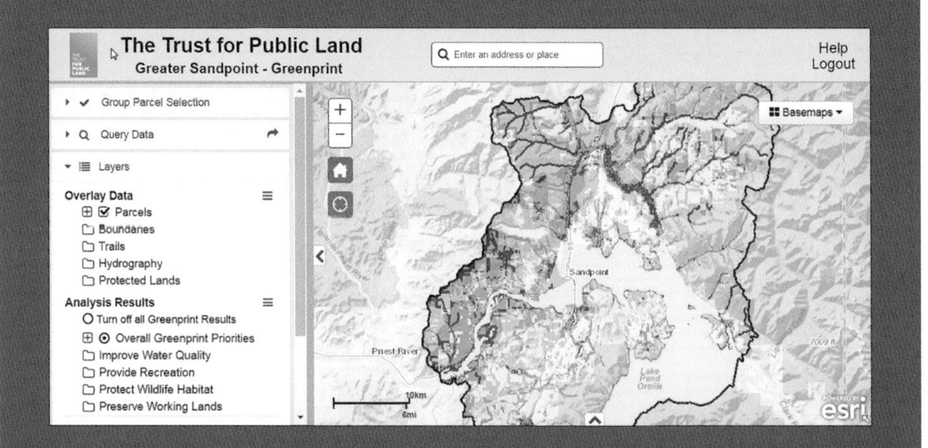

Figure 6.13. Greater Sandpoint, Idaho, Greenprint Decision Support app shows areas in red and orange that rank high for land protection based on combining water quality, recreation, wildlife habitat, and working lands data.

References

Amundsen III, Ole M. 2011. *Strategic Conservation Planning*. Washington, DC: Land Trust Alliance.

Conservation Measures Partnership. 2020. *Open Standards for the Practice of Conservation*, vol. 4.

Firehock, Karen E., and R. Andrew Walker. 2019. *Green Infrastructure: Map and Plan the Natural World with GIS*. Redlands, CA: Esri Press.

Harder, Christian, and Clint Brown, eds. 2017. *The ArcGIS Book, 10 Big Ideas about Applying the Science of Where*, second edition. Redlands, CA: Esri Press.

Jantz, Samuel M., Lilian Pintea, Janet Nackoney, and Matthew C. Hansen. 2016. "Landsat ETM+ and SRTM Data Provide Near Real-Time Monitoring of Chimpanzee (*Pan troglodytes*) Habitats in Africa." *Remote Sens.* 8 (5): 427. https://doi.org/10.3390/rs8050427.

Maryland Department of Natural Resources. 2017. *Accounting for Ecosystem Services in Charles County, Maryland*. Accessed April 2020. https://dnr.maryland.gov/ccs/Documents/Charles_Co_Ecosystem_Service_Report_Final.pdf.

TPL. 2019. "The Economic Benefits of Metroparks Toledo." Accessed February 9, 2021. https://www.tpl.org/sites/default/files/files_upload/Toledo-Report.pdf.

CHAPTER 7

Evaluating and tracking conservation impacts using data, science, and GIS

Above, migrating pronghorn deer come against a fence in Pinedale, Wyoming.

© Wendy Shattil/ Bob Rozinski, courtesy of The Trust for Public Land.

The field of medicine embraced an evidence-based approach decades ago, integrating scientific evidence, practitioner expertise, and patient preference for positive health outcomes. As a result, today's medical providers share and rely on a robust body of knowledge about treatment, prevention, and other medical approaches.

These days, a similar evidence-based approach for tracking impacts is taking root among park, conservation, and recreation organizations and agencies, but a data-driven way of evaluating impact has yet to be adopted by our field at large. Conservation and park practitioners mostly use data during planning and implementation phases for land protection and park creation but don't take full advantage of the power of data to measure effectiveness of intended impacts, cost, or adaptations.

This chapter is a call to action. It's time to go beyond using GIS to map and analyze what should be protected and to develop methods to measure success. GIS and evolving technologies take much of the expense and labor out of evaluating impacts, making this important work much more affordable and efficient. It has become easier than ever to access GIS and data, and the GIS software can be used by both professionals and people without GIS training alike. More data is now available through free and licensed sources, along with freely available data and tools from conservation organizations worldwide. The rise in coordinated partnerships, alliances, and collaboratives around the world that prioritize using GIS and spatial data for decision-making provides easier access to shared GIS data and products. Similarly, scientific research and education institutes are combining forces with land protection organizations using GIS to provide greater access to data, products, and applied research. We can now measure and monitor progress in more efficient and effective ways.

For most of our field's history, conservation organizations have relied on "bucks and acres" to capture success, summing up how many acres per year were protected and how much money was spent. The problem with this approach is that dollars and acres don't tell us if the places that are being protected are the right places for the right reasons. Are we using land protection to accomplish the goals we set out to achieve? Without GIS, science, and research, that question is almost impossible to answer. If your priority is conserving land for wildlife habitat, knowing how much land you've protected for that purpose is important, but it's just the start. We also need to know what species are benefiting from protected areas and how to monitor if that species is recovering, or still struggling, and why.

How well are conservation easements performing based on the conservation intent? How do you track and evaluate how well land protection is helping further mission goals? GIS plus technologies such as artificial intelligence and machine learning, science, and field studies aid in tracking and evaluating conservation outcomes. Advances in high-resolution data support analyses at finer scales that reveal more precise information about how well land protection is helping organizations meet mission goals. For example, high-resolution satellite and aerial imagery data reveals features such as houses, streams, trees, and roads where before only forests or dense human development were discernible. This level of detail supports better and more rapid decision-making about land and species protection, restoration, and other issues.

For instance, the Chesapeake Conservancy developed one-meter high-resolution land-cover data to support partners in the watershed to create conservation strategies for restoring properties or using best management practices to reduce sediment or pollutants in the bay. The data supports strategy development to meet the Watershed Implementation Plan goals while tracking impacts and outcomes for reporting purposes.

The examples in figures 7.1 and 7.2 show how combining high-resolution data with data collected in the field helps make the case for land protection investments. GIS supports real-time monitoring, analyzing, and reporting, with data flowing straight from the field into computer models to detect changes as they happen. Instead of acres and dollars, we can now quantify a vast spectrum of land protection benefits and impacts. We are building the evidence-based body of knowledge for conservation.

Why track and evaluate outcomes and impacts?

The reasons for tracking and evaluating outcomes are twofold: to tell the story of your organization's impacts and to combine them into big vision-driven efforts, such as the Thirty by Thirty Resolution to Save Nature and the Half-Earth Project. Quantifiable metrics add transparency and credibility to amplify your organization's impact. Metrics ensure that goals and objectives drive decision-making and support organizational health and performance. Metrics can also be combined into larger efforts to reveal how your work is contributing to a system of holistic land protection efforts.

Tell your organization's story

Your organization may already be tracking and reporting metrics beyond acres and dollars. For example, land trusts and park agencies are reporting nongeospatial information (such as volunteer hours and number of people engaging in education programs) and data calculated using GIS (such as agricultural lands being saved or improvements in water quality). This data is added to annual reports and infographics to tell the story of success. The Kansas Land Trust, for example, uses GIS

Figures 7.1 and 7.2. *Top:* The map's one-meter resolution land-cover data is for an area between downtown and the zoo in Denver, Colorado. Note the precision of the data and the ability to identify tree canopy, individual buildings, and water features. *Bottom:* This map shows the same area in Denver as figure 7.1 and displays the National Land Cover Database 30-meter land-cover product. Note how most of the features are aggregated to show development but the lower resolution makes it harder to identify specific features. Data at this resolution is difficult to use for site-level or neighborhood planning purposes.

The Chesapeake Conservancy–Conservation Innovation Center and the Denver Regional Council of Governments.

to calculate the impacts of its land protection work not only by total acres but also by miles of streams and acres of forest protected (figure 7.3).

For the parks and recreation field, GIS supports progress tracking and reporting in annual reports and updates to master plans. Infographics are a great way to distill and communicate information. In addition to using GIS to identify where new parks could be sited based on criteria such as vacant city-owned land, park systems are reporting on data-driven metrics such as green infrastructure impacts. The National Recreation and Park Association (NRPA) supports more than 60,000 parks and recreation professionals and advocates in the United States. In addition to advocacy, grants, and professional development, the organization supports the parks and recreation field with research and publications. The NRPA focuses on issues of equity, climate-readiness, and health and well-being. Figure 7.4 is an infographic that distills key information from NRPA's park metrics data for park agencies nationwide.

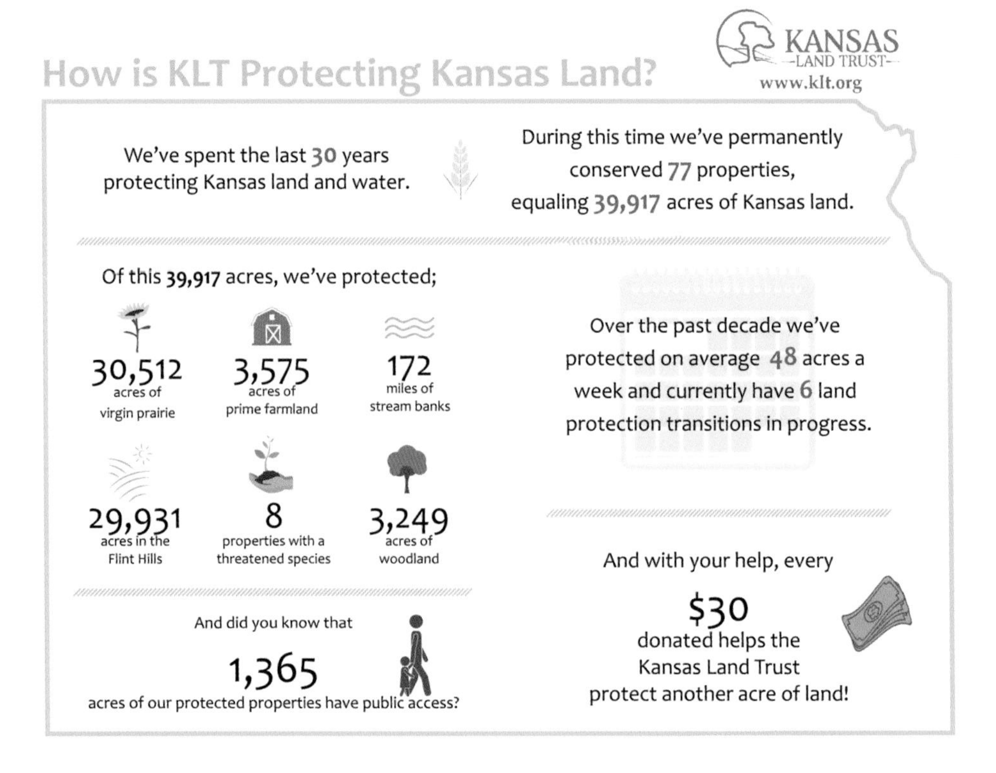

Figure 7.3. Kansas Land Trust impacts on protecting land.

Kansas Land Trust.

Are your projects working as intended?

GIS combined with new technologies helps agencies understand whether parks are working as intended. For example, if a city builds a park to absorb storm water, does it absorb enough water to prevent flooding during big storms? A city can use GIS and storm water sensors to determine whether parks are performing during big rain events to divert runoff to prevent flooding. Another question might be: Are parks being built in the neighborhoods that need them the most? If a park is designed and programmed to improve the health of the people living near it, are asthma and obesity rates declining? These are all examples of metrics that can be tracked to evaluate how well parks and protected lands are serving the needs of communities.

Improvements in automation technology are making such metrics easier to track. Measuring storm water absorption involves using special storm water gauges or monitoring equipment that requires either manual or automated data collection and reporting. Monitoring conservation easements or species movements can also be done manually or automatically. For a long time, automation technologies were fraught with barriers such as cost, infrastructure, and resources to set up, manage, process, and interpret the information. Today, the proliferation of data, online mapping apps, and packaged solutions makes automated options more attainable for organizations operating on smaller budgets and with fewer staff. The more democratized and accessible this data and these technologies are, the easier it is to build a robust science- and evidence-based culture for land protection.

Scaling up into bigger visions

In addition to technology and data, we also need alignment and coordination for authoritative and consistent methods to track conservation outcomes. Organizations analyze metrics in different ways, resulting in different outcomes and making it difficult to combine and compare results or streamline them into a consistent impact-tracking system.

The good news is that we are making progress. We have much of the data needed to drive transformational change in the way we map, analyze, and quantify the impacts of land protection. Authoritative data, such as the World Database on Protected Areas (a project between the United Nations Environment Programme [UNEP] and the International Union for Conservation of Nature [IUCN], managed by the UNEP World Conservation Monitoring Centre [UNEP-WCMC]), Protected Areas Database of the United States (US Geological Survey), National Conservation

Figure 7.4. Key findings from the 2020 NRPA Agency Performance Review.

National Recreation and Park Association, www.NRPA.org/APR.

Easement Database, Resilient and Connected Landscapes (The Nature Conservancy), Map of Biodiversity Importance (NatureServe), Global Forest Watch (World Resources Institute), ParkServe (The Trust for Public Land), and so much more data and many platforms (many of which we'll cover in the next chapter) make it possible for organizations to use and improve the same data to track impacts.

But, as described in chapter 6, we need to increase the pace, scale, alignment, and collaboration of our efforts tenfold annually if land protection is going to combat the challenges we face. How do we do that? How do we use this data to track impacts? What do we track? At the international scale, the United Nations adopted the 2030 Agenda for Sustainable Development, which provides guidance on what metrics are best for performance-based outcomes. The NRPA and the Land Trust Alliance (LTA) are providing leadership and guidance for organizations to measure and track impacts for performance and benchmarking for parks and protected lands. All organizations should join in using geospatial data and replicable and repeatable models to bring accountability and rigor to measuring impacts. It takes the guesswork out of what to protect and why and of how to measure the performance of land protection. This approach is critical to supporting the shift from reactive "emergency room" land protection to proactive and strategic conservation that will have lasting positive impacts (figure 7.5).

Figure 7.5. Frog Town Park in St. Paul, Minnesota, on October 4, 2018.

What to track?

In chapter 6, we covered how to identify conservation themes to drive land protection decisions. These themes can be a starting place to identify what to track for impacts.

Common considerations to help you decide what to track

- The conservation themes identified by your organization
- Goals of programs or initiatives
- Available funding sources for specific themes, such as water quality or habitat protection
- What others are tracking so your data can become part of a bigger effort
- The focus of collaborative research or academic partnerships, such as biodiversity data or equity

Track for funding opportunities

Available funding can be a big driver for tracking impacts and outcomes. For example, opportunities to secure federal funding from the Land and Water Conservation Fund, US farm bill, Natural Resources Conservation Service, US Department of Agriculture, and other programs are strengthened when you can quantify and map the broader impacts of your work. Equally important, data and maps can inspire your donors and supporters and show them that their money is being used effectively to create positive change.

But there's a balance between tracking too much and too little information. For example, if your organization is focused on protecting land for water quality and forest restoration, you could start with choosing one or two criteria for each of these. How do you choose these criteria?

You could start with the global framework—the United Nations Sustainable Development Goals (SDGs) and the indicators for each. These goals provide a blueprint for all countries to follow to protect the environment while building prosperity for all. Tracking land protection toward these goals recognizes conservation as an important part of the solution to create a world that balances the triple bottom line. Many SDGs are specific to conservation and include indicators that can be measured to track progress toward achieving them. Specific indicators are related to each goal so countries can measure how well they are performing against the goals. These indicators can be a guide for determining what to track and measure for your organization. An application that can help you understand how your projects relate to the SDG goals can be found at SDGtool.com. The SDG tool was developed in partnership by the University of Cambridge Conservation Research Institute, BirdLife International, Fauna & Flora International, the Royal Society for the Protection of Birds, and the International Institute for Environment and Development. It walks you step by step through a questionnaire and produces a report, in various formats, that provides graphical ways to present results and show how your projects contribute to a wide range of relevant goals and targets (figure 7.6).

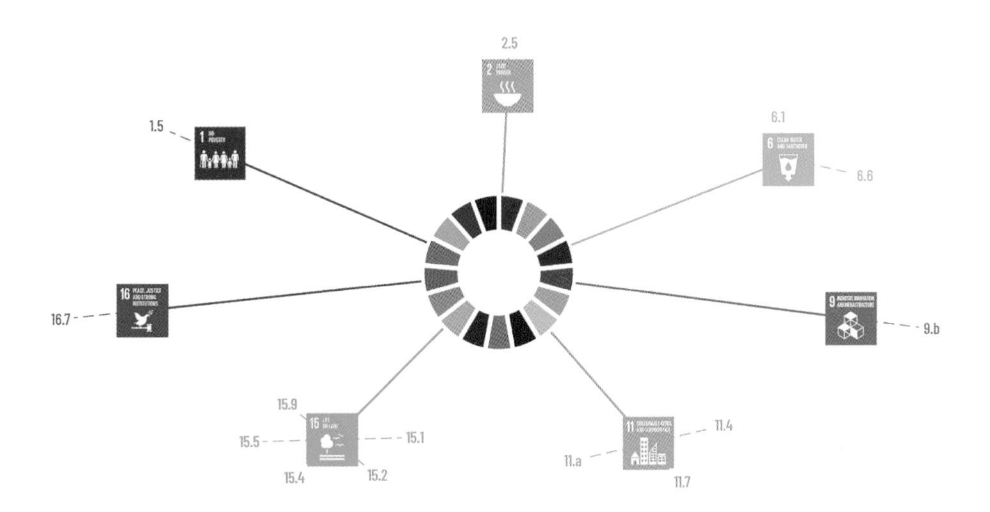

Figure 7.6. This image is from a report generated using the SDG Tool and shows the sustainable development goals, out of the United Nations' 17 SDGs, associated with a conservation project. For example, this project has relevance to the "no poverty," "life on land," and "sustainable cities and communities" SDGs. Project outcomes can be measured based on the SDG targets.

https://www.un.org/sustainabledevelopment/ [un.org]. The content of this publication has not been approved by the United Nations and does not reflect the views of the United Nations or its officials or member states.

Track for impact investing

Impact investing is becoming more popular and is a great financing mechanism for conservation and parks. Impact investing is a form of investing for social and environmental impacts versus traditional investing, which focuses on financial returns regardless of the benefits the companies or projects provide. The *Land Conservation Impact Metrics* study (Manta Consulting 2013), developed by the David and Lucile Packard Foundation, Global Impact Investing Network, Manta Consulting Inc., and the land conservation impact working group, is a good resource for impact metrics. The purpose of this study was to determine what types of metrics could best measure the performance of conservation projects for impact investors and practitioners. The metrics track back to the SDGs and use the IRIS+ impact measurement and management system.

The three-part metrics framework identified by the impact metrics study

1. The attributes of the land or the conservation values
2. The methods used to protect the lands or conservation mechanisms, which include protection, stewardship/management, or restoration
3. The benefits of conservation by measuring the conservation benefits or ecosystem services

Conservation impact investors use performance metrics to track whether their resources are going into land protection efforts that yield a high environmental, social, or economic return. In this case, understanding how land conservation projects return on the triple bottom line is the driver for high performance. Measurements related to ecology and habitat, public or human uses, and ecosystem services are considered important indicators of successful land protection projects. Some of the metrics include land type, such as streams or water bodies, habitat types, and forest tracts, all of which we can map and measure using GIS.

The LTA is another source for conservation impact metrics. The LTA performs a census every five years to determine the collective impact by land trusts nationally and at the state level. Some of the metrics that the LTA tracks are acres protected by conservation mechanisms (e.g., acquisitions or easements), how many acres of protected lands provide public access, and visitors to protected lands per year. At the state level, the LTA identifies the top three conservation priorities across all the land trusts in the state. This information can be an important indicator of what conservation themes matter the most and what should be measured and tracked statewide or nationally. For example, in New Mexico, the top three conservation priorities are protecting natural areas and wildlife habitat, protecting water and wetlands, and, tied for third, protecting farmland, forest, and scenic views. For each of these themes, criteria can be measured over time using GIS and other technologies to provide insights on how land protection is making a difference.

The NRPA identified 24 key metrics that parks and recreation agencies use to measure performance. Some of the metrics can be measured using geospatial data, such as number of residents per park. In the 2020 report, the NRPA found that, on average, there is one park per every 2,281 residents in the United States. The NRPA report includes other metrics, such as miles of trails or basketball courts or playgrounds that agencies manage. TPL developed ParkServe, which measures, for more than 14,000 cities and towns in the United States, how many people have access to parks within a 10-minute walk. It also tells how many people live within a 10-minute walk to individual parks. GIS helps us achieve different levels of granularity and scale, depending on our needs and goals. This data provides important benchmarks that park agencies can use to determine where they are excelling or possibly falling short on providing the right park system benefits in their communities.

Common conservation and park themes to track

- Acres
- Dollars
- Habitat types (i.e., waterfowl habitat, pollinator habitat)
- Forest cover
- Stream miles protected
- Headwaters protected
- Trail miles created
- Increased access to public lands
- Park and public land use
- Economic benefits of conserved lands or parks

- Ecosystem services
- Park amenities
- And much more

You've chosen what to track—now what?

Steps for identifying data, analysis, and products to support tracking and evaluating outcomes

1. Consider these important questions: Does the data exist to track what you've chosen? And if the data exists, is it at the right scale, resolution, or currency to provide defensible results? Is the data updated frequently enough to support your reporting needs? Does the data come from an authoritative source? Will the data schema or attributes change in a way that will create issues for rerunning models or comparing results? If data doesn't exist, how might you either create it or use proxy data? Your answers to these questions will determine how and what you track.
2. Create a replicable analysis. Models and Python scripts are great for this task. When rerunning the analyses for the purposes of measuring impacts and outcomes, the steps and processes will likely remain the same. This methodology supports defensible results and provides data integrity and rigor for reporting, research, and scientific uses.
3. Determine who will be using the information and how. Will this data mainly be used for an annual report, an infographic, a website, or social media? Will this data be used for scientific research? Peer-reviewed journal articles? Funders? Knowing the uses will help you understand how to capture metadata and methodology

information and how to model the data for the results and attributes that may be needed for purposes such as scientific studies.

4. What products will you use to tell the story of your impacts? Infographics such as the ones earlier in this chapter are great for graphically synthesizing data for reports and websites. ArcGIS StoryMaps provides online, interactive web apps for impact reporting and visualizing. For example, the StoryMaps swipe function can be a powerful way to show before and after results from land protection. The images from Ducks Unlimited Canada (DUC) in figures 7.7 and 7.8 show the before and after impacts on waterfowl habitat of draining wetlands in the North American

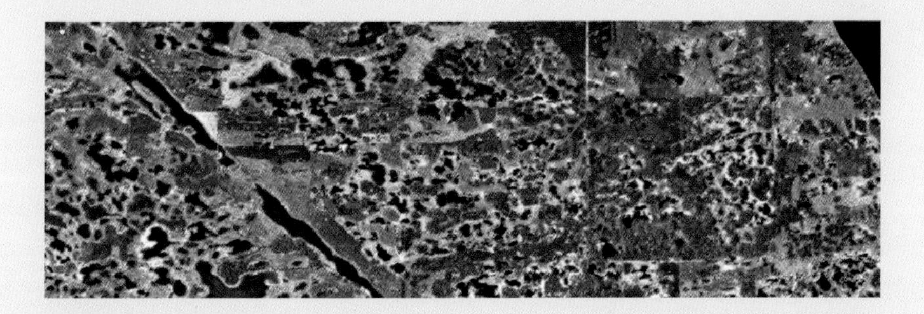

Figure 7.7. Before—intact wetlands in the prairie pothole region of North America.

Credit Ducks Unlimited Canada.

Figure 7.8. After—the same image extent after wetlands were drained for agriculture.

Credit Ducks Unlimited Canada.

prairie pothole region. DUC used the swipe function on its website so users could interact with the before and after images. In addition, DUC uses GIS to calculate statistics about the impacts that land protection, restoration, and management decisions have on waterfowl habitat. Powerful before and after images such as these show why it's important to track conservation impacts and how the land has changed to better understand what we can do to restore the important functions it once served.

Monitoring conservation properties for impacts

Land trusts are experts at assessing the current state and impacts of protected, privately owned lands. When land trusts hold conservation easements on these lands, they are legally required to monitor the properties to ensure they are being managed for conservation and that there are no violations, such as new development or land-use changes. In its *Land Trust Standards and Practices* report, the LTA (2017) provides guidance on what is required by land trusts for conservation easement monitoring. Each easement differs on the basis of the features of the property and the intent of the landowner, and monitoring criteria vary accordingly. Some easements protect high-value habitat, forests, or working agricultural lands; others provide recreation access. Not all easements are on private lands; some provide additional protections on publicly owned lands (e.g., protections for the desert tortoise on military bases). Monitoring easements is an important form of tracking and measuring the impacts of land protection.

Methods for monitoring properties vary. Traditionally, land trust staff would perform site visits, taking notes with pen and paper and pictures of the property. Although site visits are still important for land trusts to maintain their relationships with landowners and the land, they're now being complemented by more advanced technological monitoring methods. Smartphones or handheld GPS units capture pictures, coordinates, and survey questions and data.

Increasingly, land trusts are using more sophisticated technology to complement site visits for monitoring. The use of high-resolution satellite imagery and machine learning to detect land-use/land-cover changes on properties is growing in popularity. The LTA has many resources on its website that give an overview of the methods land trusts are using for monitoring sites. The University of California, Berkeley, produced a report on the remote monitoring of conservation easements, providing information on the ways land trusts are using remote sensing products (from satellites and aircraft) and the benefits, challenges, and best practices. Companies are providing solutions for land trusts and conservation organizations to remotely monitor conservation properties. Esri offers levels of access to imagery that spans a range of spatial resolution, temporal coverage, and cost. The US Department of Agriculture's National Agricultural Imagery Program imagery, freely available through ArcGIS Living Atlas of the World, is used by organizations to detect land-cover and land-use change. You can compare and analyze this imagery in ArcGIS Online with little training or background. Many data and technology approaches are available to conservation organizations monitoring protected places. Pick the one that best fits your organization's requirements and resources.

Monitoring helps ensure that protected places remain true to the original conservation intent. But to achieve the needed level of protected land, we are going to have to expand beyond monitoring easements for compliance to conservation requirements to monitoring easements for conservation impacts, such as improving water quality or habitat using data-driven methods.

From planning and implementing to tracking and managing

The emerging conservation paradigm combines data collection, management, analysis, and storytelling to support tracking and management of protected places. Access to ArcGIS® Solutions for Conservation and new combinations of technologies are lowering the barriers to near-real-time tracking and management. In the past, these types of solutions were too expensive and complex for many organizations to adopt. ArcGIS Online, which is affordably licensed through the Esri Nonprofit Organization Program or the Esri Conservation Program, makes it easier for organizations of all sizes and resources to access ArcGIS Enterprise functionality.

A powerful example is ArcGIS Solutions for Conservation, a set of apps and workflows to help protected area managers go beyond using GIS for planning and implementation. These apps and workflows support tracking, capturing, and managing information for better protected area management and for tracking conservation impacts. These out-of-the-box solutions lower the barriers, both perceived and real, that inhibit organizations from using the full capacity of the technology and software available to them. Organizations such as African People & Wildlife (AFW) and parks such as Garamba National Park in the northeastern Democratic Republic of the Congo are using these GIS solutions to transform the way they operate by accessing real-time data and apps. They are literally able to save the lives of people and wildlife through data-driven insights and interventions (figure 7.9).

Esri is working with conservation leaders worldwide to build this suite of apps and workflows to support better conservation impacts and faster outcomes. These solutions are transforming the way managers of protected lands use data to make key management decisions in alignment with community needs and conservation intent. GIS has transformed our efforts to protect land. Now it's transforming our ability to track, manage, and measure outcomes and pivot when needed. Change is happening on our planet faster than we could ever imagine. Technology has been part of the problem, and now it's a core part of the solution.

ArcGIS® Solutions for Conservation apps

At the time of writing this book, ArcGIS for Conservation supported the following apps:

- Protection Operations—a collection of maps and apps used by protected area staff to track poaching or illegal activity and monitor the status of protection operations within and around protected areas.
- Wildlife Management—a collection of maps and apps used by protected area staff to capture wildlife observations and monitor the status of wildlife populations within and around protected areas.
- Conservation Outreach—a collection of maps and apps used by protected area staff to manage wildlife conflicts and monitor the status of outreach activities within and around protected areas.

Figure 7.9. African People & Wildlife team members using GIS.

African People & Wildlife/Felipe Rodriguez.

References

LTA. 2017. *Land Trust Standards and Practices*. Washington, DC: Land Trust Alliance.

Manta Consulting. 2013. *Land Conservation Impact Metrics*. http://www.mantaconsultinginc.com/images/reports/Land-Conservation-Metrics.pdf.

CHAPTER 8
Online data and tools for conservation and park efforts

Above, view of Zion National Park from Zion's Kolob Terrace on the western side of the park, near Tabernacle Dome.

© Mike Schirf, courtesy of The Trust for Public Land.

The ways we access and use data and geospatial technologies have evolved dramatically over the last decade. The biggest shift has been from individual desktop to cloud-based systems and applications, a shift that democratizes geodata and information to power the simplest to the most complex projects and applications.

You can access many online applications and datasets free of charge or by subscription. Sources range across nonprofit organizations, for-profit companies, agencies, and academic institutions. Sometimes you don't need to invest in GIS capacity if you have a discrete and targeted use for data that already exists in an online mapping app. You can also use online data and applications to enhance or complement the GIS work you are already doing. For example, you might be using GIS to create maps and perform spatial analysis, but you don't have the expertise on staff to do more complex modeling, big data analytics, or scientific research. In this case, data that you need may already exist. You can download or connect to data from other sources to integrate directly into your GIS project or into your web app. If the data you need doesn't exist, you can explore partnership opportunities with research or academic institutes or organizations, or hire a consultant or company to perform the more complex analytics. Following are examples of how to use freely available online data and tools for conservation and park issues.

Use case 1: Climate resilience and biodiversity

In this case, you need to demonstrate that a property your organization can protect will be a critical missing link for climate-resilient and connected wildlife habitat as the climate changes. You are in luck. Top scientists and technical experts at The Nature Conservancy (TNC) have developed the Resilient Land Mapping Tool web app. Use this web app to create a map that shows the climate significance of a property to strengthen the case for funding and acquisition. Zoom into the area of the project, and then either upload the parcel boundary or sketch a polygon to represent the parcel and click a button to run a report (figure 8.1).

The Resilient Land Summary report provides rich information about resilience, landscape connectedness, and local connectedness (figure 8.2).

You can also use this tool to identify areas where your organization might want to work to protect lands for resilience and connectedness. By turning on the secured areas and the resilient and connected network layers, you can easily see where areas are already protected and where areas critical to resilience and connectivity are not protected. You can explore the map, upload your own places of interest including parcels, sketch polygons of interest, and use this tool to inform where public or private lands managed for conservation can make a big impact (figures 8.3 and 8.4).

Download TNC Resilient Sites data to add to maps in ArcGIS Pro or ArcMap. In the Resilient Land Mapping Tool web app, you can also add an ArcGIS map service to integrate other data that either your organization creates and maintains or that comes from other organizations, agencies, or companies.

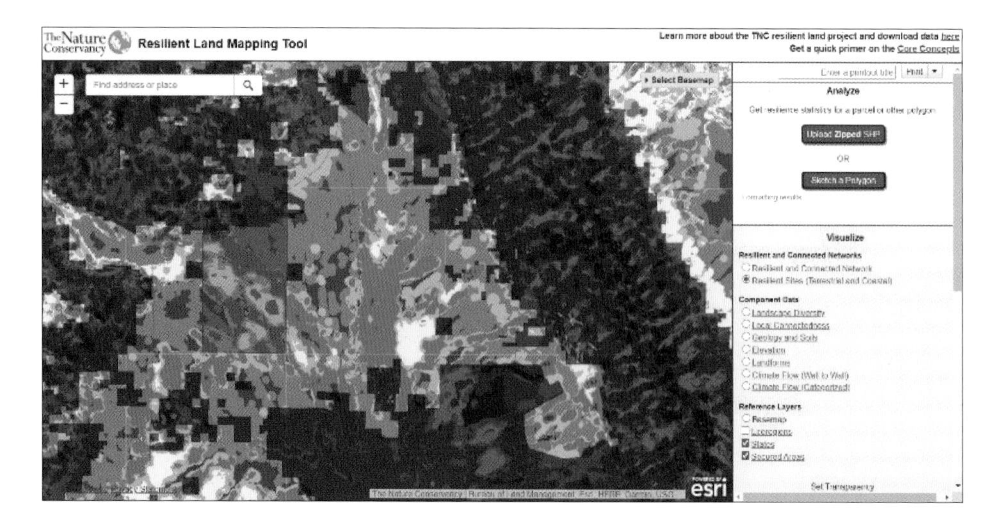

Figure 8.1. Map showing protected lands in blue, terrestrial and resilient sites data, and a highlighted land protection potential property in red. Map created using the TNC Resilient Land Mapping Tool app.

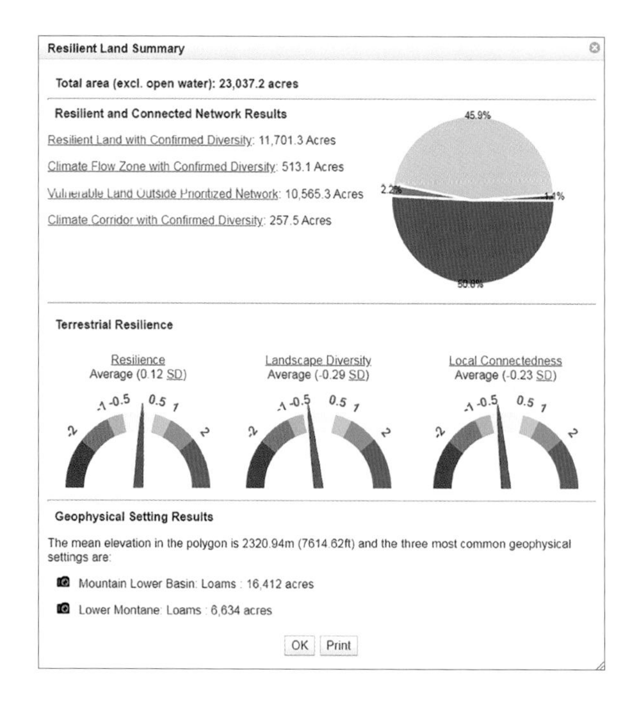

Figure 8.2. Resilient Land Summary from the TNC Resilient Land Mapping Tool app.

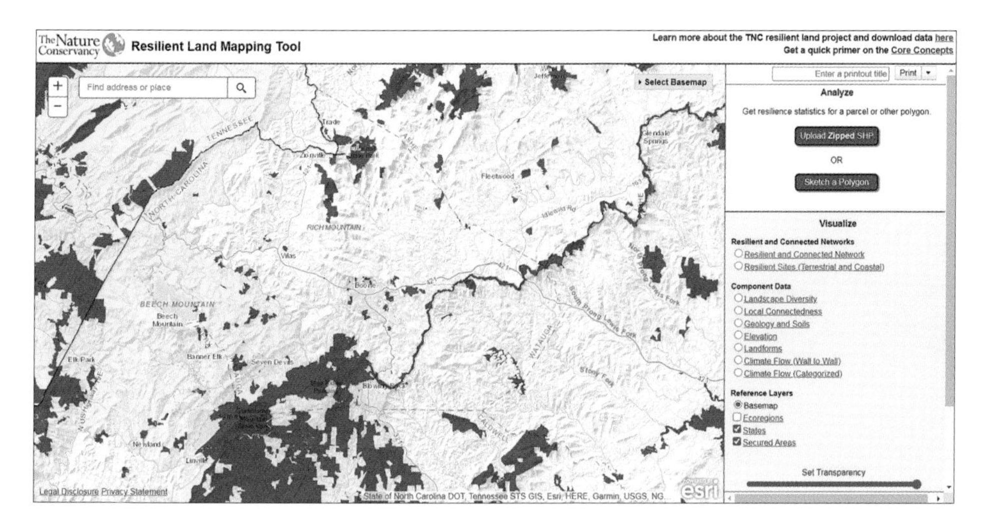

Figure 8.3. This map shows areas in blue (secured areas) that are protected in the northwest area of North Carolina, including Boone and Blowing Rock. Map created using the TNC Resilient Land Mapping Tool app.

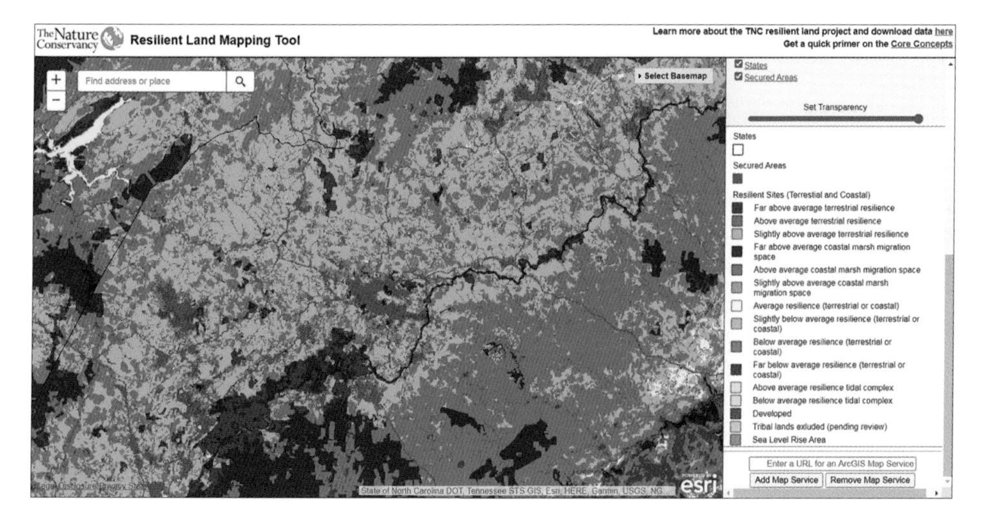

Figure 8.4. This map shows the same area in northwest North Carolina as in figure 8.3, with the Resilient Sites layer overlaid with the secured (protected) areas layer in blue. Resilient sites include places that will retain high-quality habitat and support species as the climate changes as these places have complex topography and connected land cover. Map created using the TNC Resilient Land Mapping Tool app.

To further strengthen your case with species-specific data, use NatureServe's Map of Biodiversity Importance (MoBi) app and data. This app enables you to create maps from an array of species data, including imperiled species. As with TNC data, you can download this data to create your own maps or access it through ArcGIS Online and ArcGIS Living Atlas of the World. When working locally, search for data from sources in the areas you are analyzing or mapping. For example, the North Carolina Natural Heritage Program has an online web mapping app and data for download. This data may be at a higher resolution and scale that supports finding areas of importance for biodiversity protection in North Carolina. Experts in these organizations have created these products and made them freely available to democratize information to fuel an increase in land protection focused on climate resilience and biodiversity. By accessing the online web mapping apps, you can create maps to include in reports or documents, or in your own GIS to support the protection of these places.

Use case 2: Making the case for park development with budget constraints

In this case, you work for a park agency facing budget cuts, but the agency has several parks in the queue for development. The land is already secured and the community has already provided input on the final park designs. Your task is to use GIS and data to support the park department in its efforts to convince the city council that continuing with constructing and opening these parks needs to remain a top priority as they are going through the budgeting process. These parks are all located in high park need areas of the city. The neighborhoods desperately want and need these parks and need data and maps to advocate for the parks to their city council members.

The ParkServe app from The Trust for Public Land (TPL) is a free, authoritative resource for this purpose. The organization's GIS and park experts combined research on park access with sophisticated data and GIS modeling approaches, including tools such as ArcGIS Network Analyst, to create this app. It offers statistics and maps that show how many people in your city have access to a park within a 10-minute walk and where parks are needed most. Areas of park need are ranked from low to high. By using this information, you can show how these potential parks fit within the park system of the city and could fulfill park needs. You can report on the socioeconomic characteristics of neighborhoods served by these parks. TPL includes urban heat island data to further your case. In addition to using the online app, you can download ParkServe data to create your own maps and do your own analysis in ArcGIS Pro. The data is also available through the US Geological Survey (USGS) Protected Areas Database of the United States (PAD-US) product and ArcGIS Living Atlas, accessible in ArcGIS Online.

The city budget situation also might allow only one park to be developed in the current fiscal year. ParkServe supports geodesign. With the ParkServe Park Evaluator tool, you can create reports that include how many new residents will be served by each park, along with the socioeconomic, race, and ethnicity characteristics of these people. This information can help the city rank each park on the basis of the highest benefit it will provide and select the park to be developed based on an objective, data-driven process. (Note: There are many other data layers and factors that go into decision-making. The maps in this example are just one instance of how data can inform a decision-making process.) This information also supports the city and the park department in exploring alternative funding options from nontraditional partners such as health-care providers, for-profit companies, or foundations. Or you can use it to make the case for a public park ballot measure in which residents decide to create a new tax or redirect taxes already collected from sources such as recreation equipment sales or sports gambling to a park development, operations, and maintenance fund. There are many good examples in the United States where these types of measures have passed and funded new parks, trails, and protected lands. Information on park ballot measures that have passed can be found at TPL's free online resource LandVote. TPL's Conservation Almanac is another free online resource that provides information on funding programs and sources for parks and protected lands.

Use case 3: Using GIS to support restoration activities for a landowner

In this case, the governor of your state has just issued a commitment to reduce sediment buildup for rivers. A conservation landowner who has property adjacent to a river has contacted you. She wants to know what restoration measures she can take to contribute to the reduction of sediment into the river, as well as to the conservation goals for the whole watershed. You don't have GIS at your land trust but you want to support her in her efforts, and there is a free online mapping app available that you can use to help her (figure 8.5).

This example features an online mapping app specific to a subset of counties in Pennsylvania, but tools such as this may exist in your area. Search online to determine what is available for discrete efforts such as this. Restoration Reports is a free tool developed by Chesapeake Conservancy, a Chesapeake watershed-focused land trust, to support restoration. This free, online mapping app is powered by high-resolution land-use/land-cover data generated by the conservancy across five Pennsylvania counties. It identifies restoration activities that can contribute to improving specific ecological and environmental conditions. Landowners, land trusts, or agencies can use the tool to create reports for properties in these five counties. After entering an address, the tool returns information on the composition of the property calculated using one-meter high-resolution land-cover data. Data returned includes a breakout of land characteristics, such as tree canopy cover, impervious surfaces, and water features.

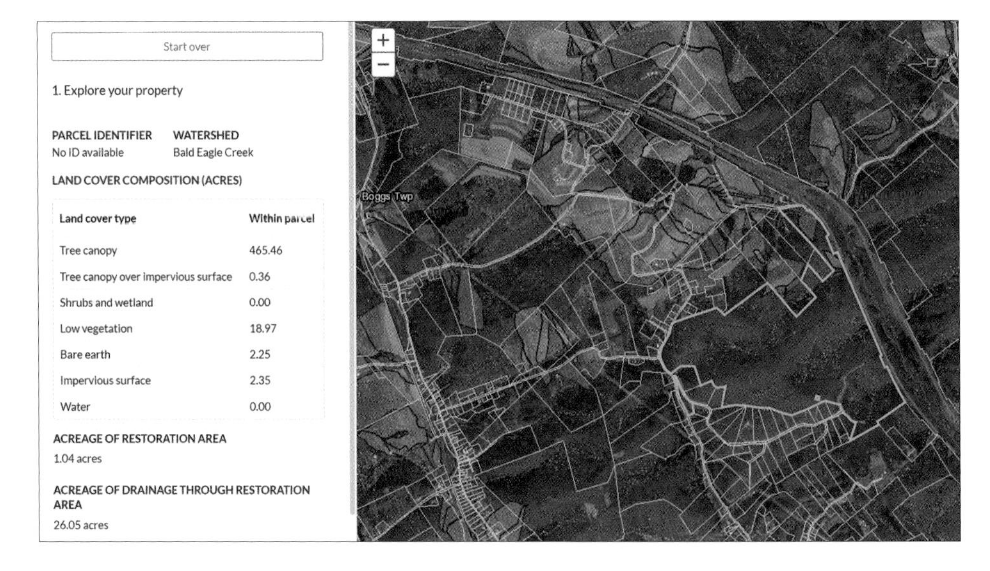

Figure 8.5. Restoration Reports online mapping app produces a report that provides private landowners with data, maps, and information about the restoration opportunities on their property. The map shows land parcels symbolized with yellow lines, restoration areas in pink, and drainage in blue. The report provides a breakout of land-cover type by acre per parcel along with acres of drainage and restoration areas.

The Chesapeake Conservancy.

To generate the report, the user selects from the three management priorities: agriculture, wildlife, and recreation. In this case, we'll choose all three because reducing sediment loads into the river positively affects each one. The report includes recommendations for restoration tactics to improve water quality on the property and for the river the waters flow into. For example, by focusing on slowing water in the identified flow paths on the property and planting trees in specific locations, sedimentation can be slowed or stopped before it enters the river. In addition, the report provides educational information for the landowner on the multiple benefits that these restoration efforts will provide, such as enhancing wildlife habitat and improving air quality. The report also links the landowner to different conservation professionals in the field who can further help and support her restoration efforts (figure 8.6).

Free online tools such as this can greatly enhance the pace and scale of restoration efforts from the local to the larger watershed or landscape scale.

Use case 4: Preventing pipeline development on protected lands

In this case, your land trust manages conservation easements that are in a proposed pipeline development zone. These easements protect intact forest and water resources and are part of an important wildlife corridor. A pipeline company is trying to make the case to the county commissioners to make an exemption to the development restrictions on the easements so the pipeline can be constructed through the originally planned route. You need to make the case that this development would severely impair the conservation qualities that these lands provide as a public good. There are other routes that the pipeline development can follow. You need a map to show that the pipeline construction can entirely avoid these easements with minimal impacts to the community and the company.

The National Conservation Easement Database (NCED) is a free, online resource that includes an interactive map of easements in the United States. You can view easements; get information about the easements, such as who owns and manages them; use drawing tools to illustrate on the map; and print or export maps. It's a free and easy resource to create maps or map images that you can use to make the case for maintaining the original conservation intent of private conservation easements. NCED also offers the ability to produce PDF reports for individual

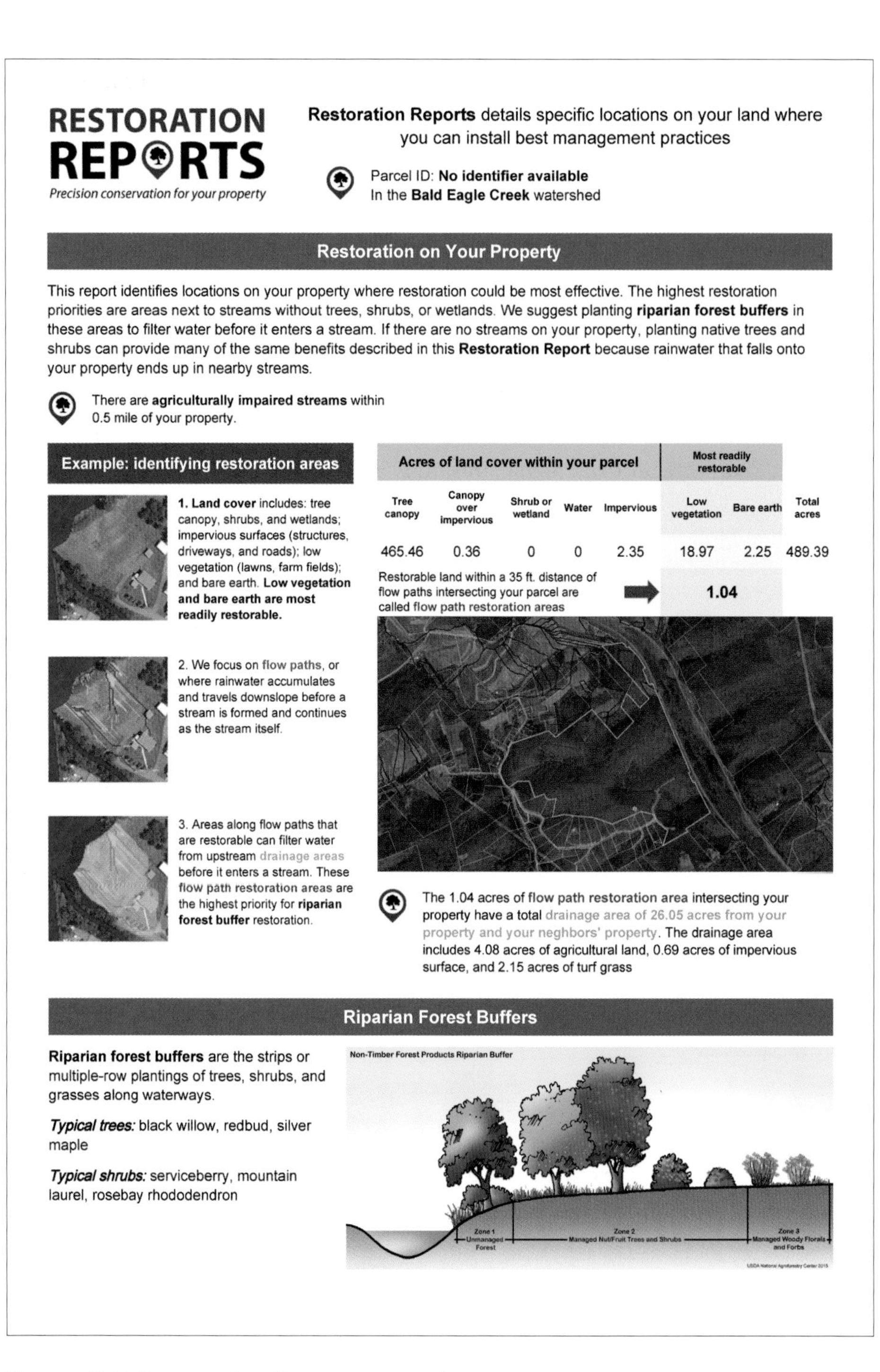

RESTORATION REPORTS
Precision conservation for your property

Restoration Reports details specific locations on your land where you can install best management practices

Parcel ID: **No identifier available**
In the **Bald Eagle Creek** watershed

Restoration on Your Property

This report identifies locations on your property where restoration could be most effective. The highest restoration priorities are areas next to streams without trees, shrubs, or wetlands. We suggest planting **riparian forest buffers** in these areas to filter water before it enters a stream. If there are no streams on your property, planting native trees and shrubs can provide many of the same benefits described in this **Restoration Report** because rainwater that falls onto your property ends up in nearby streams.

There are **agriculturally impaired streams** within 0.5 mile of your property.

Example: identifying restoration areas

1. Land cover includes: tree canopy, shrubs, and wetlands; impervious surfaces (structures, driveways, and roads); low vegetation (lawns, farm fields); and bare earth. **Low vegetation and bare earth are most readily restorable.**

2. We focus on flow paths, or where rainwater accumulates and travels downslope before a stream is formed and continues as the stream itself.

3. Areas along flow paths that are restorable can filter water from upstream drainage areas before it enters a stream. These flow path restoration areas are the highest priority for **riparian forest buffer** restoration.

Acres of land cover within your parcel					Most readily restorable		
Tree canopy	Canopy over impervious	Shrub or wetland	Water	Impervious	Low vegetation	Bare earth	Total acres
465.46	0.36	0	0	2.35	18.97	2.25	489.39

Restorable land within a 35 ft. distance of flow paths intersecting your parcel are called flow path restoration areas → **1.04**

The 1.04 acres of flow path restoration area intersecting your property have a total drainage area of 26.05 acres from your property and your neghbors' property. The drainage area includes 4.08 acres of agricultural land, 0.69 acres of impervious surface, and 2.15 acres of turf grass

Riparian Forest Buffers

Riparian forest buffers are the strips or multiple-row plantings of trees, shrubs, and grasses along waterways.

Typical trees: black willow, redbud, silver maple

Typical shrubs: serviceberry, mountain laurel, rosebay rhododendron

Figure 8.6. Restoration Reports example.

Chesapeake Conservancy.

easements, query for easements managed by land trusts, and download the data for use in GIS or other applications. Both public and private organizations use NCED data for smart conservation and development planning. Having access to this information up front can help alleviate expensive legal costs that may arise because of incompatible uses. The Land Trust Alliance has leveraged NCED data to successfully improve federal energy policies that favor the protection of easements and the use of easements as mitigation tools to offset the negative impacts of energy development. This information is readily available for anyone to use (figure 8.7).

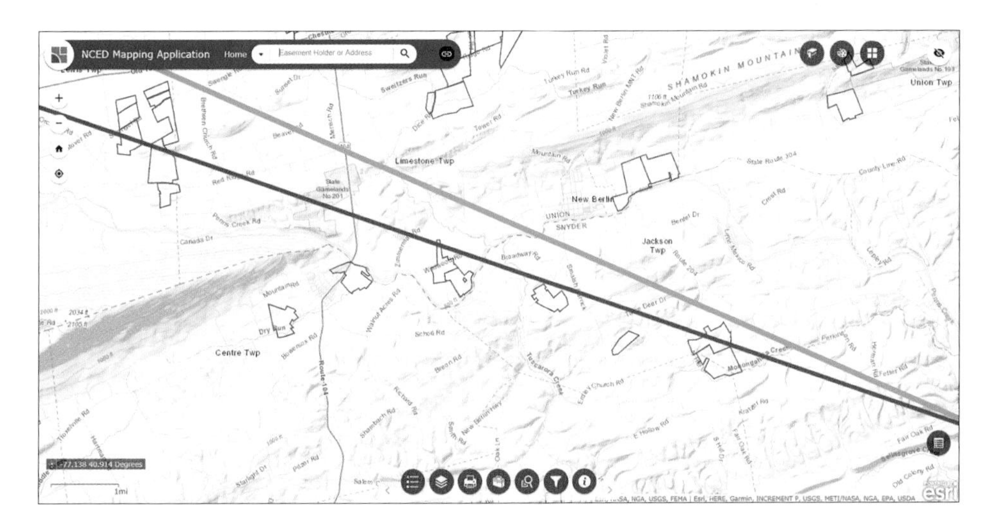

Figure 8.7. This fictional example map was created using the NCED online interactive map. The red line represents the fictional proposed pipeline intersecting easements and the green line represents the fictional proposed alternative route for the pipeline that does not cross through existing easements.

National Conservation Easement Database.

These are just a few examples of uses for free online mapping tools and data. You can also access online mapping resources through a paid subscription or license. In this chapter, I'll provide a brief overview of some of the many resources available to the conservation and park community. Of course, it's also worth doing internet searches for online mapping tools and data using key search terms that match your goals.

Free online resources from organizations

Many nonprofit organizations provide freely available data and tools through online websites and interactive applications. Following are a few examples.

Chesapeake Bay Watershed Land Cover Viewer

The viewer, developed by the Chesapeake Conservancy, enables you to explore one-meter high-resolution data in the watershed. The swipe function allows you to move between high-resolution imagery and the data to visualize land cover in more detail. For example, the imagery enables you to see what type of structures the land-cover category represents. You can also download the data for use in your own GIS.

Last Chance Ecosystems

The Last Chance Ecosystems web mapping app provides access to several datasets created by TNC scientists. Some of these layers include last-chance crisis ecosystems, last-chance mammal areas, and crisis ecosystems, as well as the resilient and connected network data referred to earlier in the chapter. This data displays at varying scales as you zoom in and out and provides guidance on where to focus land protection efforts that are most critical to the protection of species and habitats in the face of human development and climate threats. Last Chance Ecosystems is one of the many free, online web mapping apps available through TNC's Geospatial Conservation Atlas.

ProtectedLands.net

GreenInfo Network's applications showcase the USGS PAD-US database through an interactive map app called protectedlands.net. You can explore public lands by attributes such as manager name, manager type, biodiversity protection, and protection mechanism. You can also search for protected lands by name and access stories about how agencies and others are using the PAD-US data for conservation planning and land management purposes. GreenInfo Network has an array of mapping apps you can access through its website, many of which are specific to the state of California.

Conservation Almanac

The Conservation Almanac, developed by TPL, includes an interactive map that tracks conservation spending in the United States since 1998. Spending is tracked at the federal, state, and local levels. The map shows the boundaries of protected lands that you can click on to get spending information in addition to attributes such as purchase type, year the deal closed, and the property manager. You can query by county, purchase type, or spending program and have the option to produce a PDF report for individual properties or a Microsoft Excel table for aggregated query results.

ParkServe

ParkServe, developed by TPL, was covered in chapter 2 but is worth noting here again as a good tool to map and understand park access and need. As described in use case 3 earlier in the chapter, ParkServe can also evaluate new park development by allowing you to sketch potential parks.

LandScope America

The LandScope America website, developed by NatureServe and the National Geographic Society, combines conservation science, information technology, and multimedia storytelling to inspire support for conservation. The web mapping app includes data on themes such as conservation priorities, threats to conservation, and ecosystems. You can explore data, save and share maps, and create comprehensive overview reports by state, watershed, or county.

Global Forest Watch

Global Forest Watch (GFW) is an initiative of the World Resources Institute in collaboration with many for-profit companies, nonprofits, academic and research institutes, and agencies. GFW is an online mapping platform with data and tools to track and monitor forests worldwide. You can access data on deforestation, fires, climate, specific land uses, and biodiversity. You can analyze changes in forest, land use, and biodiversity using online analysis tools and access the most up-to-date satellite imagery. Detailed reports, a dashboard, and the Forest Watcher app discussed in chapter 5 are also included (figure 8.8).

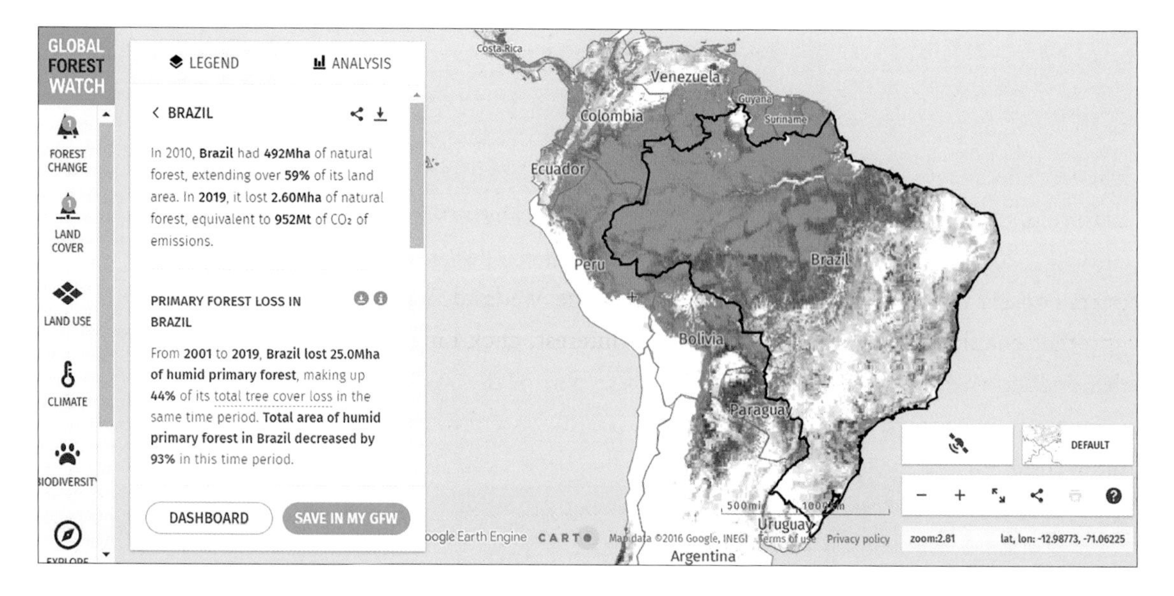

Figure 8.8. World Resources Institute Global Forest Watch interaction mapping app.

Global Forest Watch and World Resources Institute.

Important Bird Areas

The National Audubon Society provides the interactive map Important Bird Areas. You can explore continental, global, or state bird priority areas. Clicking on the areas reveals information on the name of the area, the scale of importance, and the local Audubon chapter that manages that area. Local chapters such as Audubon California have created their own focused interactive web map.

Free online resources provided by US government agencies

Governments all over the world manage data and tools that they make available to the public. Following are just a few examples of freely available geospatial resources that include data and tools for mapping and analyzing issues related to protected areas.

Wetlands Mapper for National Wetlands Inventory

The US Fish and Wildlife Service maintains an inventory of wetlands in the United States, the National Wetlands Inventory. The data includes information on the type of wetlands, distribution, surface water habitats, and other information that supports conservation efforts. The agency created the Wetlands Mapper, a mapping app that enables you to zoom to an area of interest, click on a wetland, and get detailed information about the size and habitat type of the wetland. You can also view historic mapping reports related to the wetland and download the data from the website for use in your own apps.

EnviroAtlas

The US Environmental Protection Agency provides a robust mapping tool that includes more than 400 data layers related to ecosystem services, the benefits that humans get from nature. The data is organized in the following categories: clean air, clean and plentiful water, natural hazard mitigation, climate stabilization, recreation, culture and aesthetics, food, fuel and materials, and biodiversity conservation. Some of the types of data included are carbon storage by trees, acres of pollinator crops with no nearby pollinator habitat, birdwatching recreation demand, daily domestic water use, and much more, along with analysis tools for climate change and many others. You can save your session to come back to it later and mark up and print maps.

The National Map

The USGS has created a mapping application to provide access to topographic data. Some of the datasets include historic and current topographic maps, US boundaries, elevation products, land cover, hydrography, National Agricultural Imagery Product, and elevation data, among many other data layers. There is a link on the map to download the data. You can add data from other agencies such as the USGS PAD-US or Federal Emergency Management Agency flood boundaries. You can also mark up and print maps from this website.

PAD-US

PAD-US, managed and published by the USGS Gap Analysis Project, is the authoritative, comprehensive database of protected lands in the United States. The data includes fee lands, overlapping designations, easements, proclamation boundaries, and marine areas. It also includes an array of attributes about land manager, protection mechanism, and more. PAD-US is available for download and can be accessed through online mapping apps such as ArcGIS Online, ArcGIS Living Atlas, and The National Map.

LANDFIRE

The US Department of Agriculture (USDA) Forest Service and Department of Interior manage and publish LANDFIRE, a resource for data, information, and tools. The website includes a rich set of vegetation and wildland fire/fuels data to support fire and resource management, planning, and analysis. There is a data distribution tool and instructions on how to bring the data into ArcGIS Online or ArcGIS Pro.

National Water and Climate Center

The USDA National Resources Conservation Service provides an interactive map through its National Water and Climate Center website to explore data on snow and climate monitoring and water supply. Datasets include snow depth, snowpack, soil moisture, frequencies, and real-time data streams from SNOTEL (automated snow monitoring) and other real-time data sources. You can view data by station or basin.

Subscription online resources

Many companies offer data and tools through subscription services. They charge a monthly or annual fee to access data and online analysis capabilities. These services offer access to a range of data and tools of varying complexity, with different levels of subscription that unlock access to more complex data and more tools to analyze, model, and interpret this data. Following, I'll provide an overview of a few of the many companies that provide subscription access to geospatial data and tools.

ArcGIS Online

Through ArcGIS Online organizational accounts, you have access to online mapping, along with analysis tools that summarize or enrich data and analyze patterns and proximities. Certain subscription levels allow access to services such as ArcGIS® GeoEnrichment Service, which provides the ability to get facts and data about a geography of interest or within a certain distance or drive time of a location. Using ArcGIS Online, you can also access ArcGIS® Business Analyst, through an Esri subscription service, to get socioeconomic data and thousands of other third-party datasets to help you answer questions about the community you are analyzing. Businesses use the tools to analyze questions such as where to site a new store. Conservation and park professionals can use it to site parks, trails, and open spaces. They can also run reports on topics such as consumer spending; for instance, where are community members purchasing outdoor equipment? Understanding purchasing habits and behaviors, in addition to demographic and socioeconomic data, can give you insights into where parks and open space are needed most. You can even access traffic count data to understand where to put an entrance to a park or open space.

Descartes Labs platform

Descartes Labs offers subscription access to a platform that includes a vast repository of data and an artificial intelligence (AI)-based technology stack to help data scientists answer complex questions fast. TPL used the Descartes Labs platform to create the high-resolution urban heat island dataset for the United States in a fraction of the time it took in other computing environments.

Upstream Tech Lens

Upstream Tech Lens offers subscription web-based access to a planning and monitoring platform that combines computer science, machine learning, and remote sensing. Many land trusts are using this platform for remote conservation easement monitoring and to oversee and track restoration progress.

Conserve.io

Conserve.io combines cloud-based and mobile technology with big data into a platform to address conservation issues. The current apps focus on marine issues, such as helping boaters know where they risk hitting manatees. Another app provides information to reduce whale and ship collisions, while yet another tracks great white shark sightings to reduce human/shark encounters and promote a peaceful coexistence. The blending of crowdsourced data and technology is helping conservation and regulatory professionals and the public protect important species.

Integrated Biodiversity Assessment Tool

The Integrated Biodiversity Assessment Tool (IBAT) is a web-based mapping and reporting tool, developed and maintained by an alliance of global environmental organizations (BirdLife International, Conservation International, United Nations Environment Programme—World Conservation Monitoring Centre [UNEP—WCMC], and the International Union for Conservation of Nature [IUCN]). IBAT offers commercial access to globally authoritative data on biodiversity and protected areas from three sources: IUCN Red List of Threatened Species, World Database on Protected Areas, and World Database of Key Biodiversity Areas. The goal of the alliance is to provide the most up-to-date science and data to inform decision-making regarding sustainable development and investments from the private sector while reducing the impacts on biodiversity (figure 8.9).

For no fee, you can access a data map to view and explore the datasets. You can also access high-level biodiversity reports by country, including information such as extinction risk assessments, the red list index, and threats by species and taxonomic groups (figure 8.10).

Subscription options include access to site-specific reports that include more in-depth information, plus GIS downloads. Both the free and subscription options offer the ability to save projects and reports in a dashboard. This is a good example of a business model, supported by subscriptions, that maintains and updates important protected lands and biodiversity data while also providing high-quality free access to data and reports.

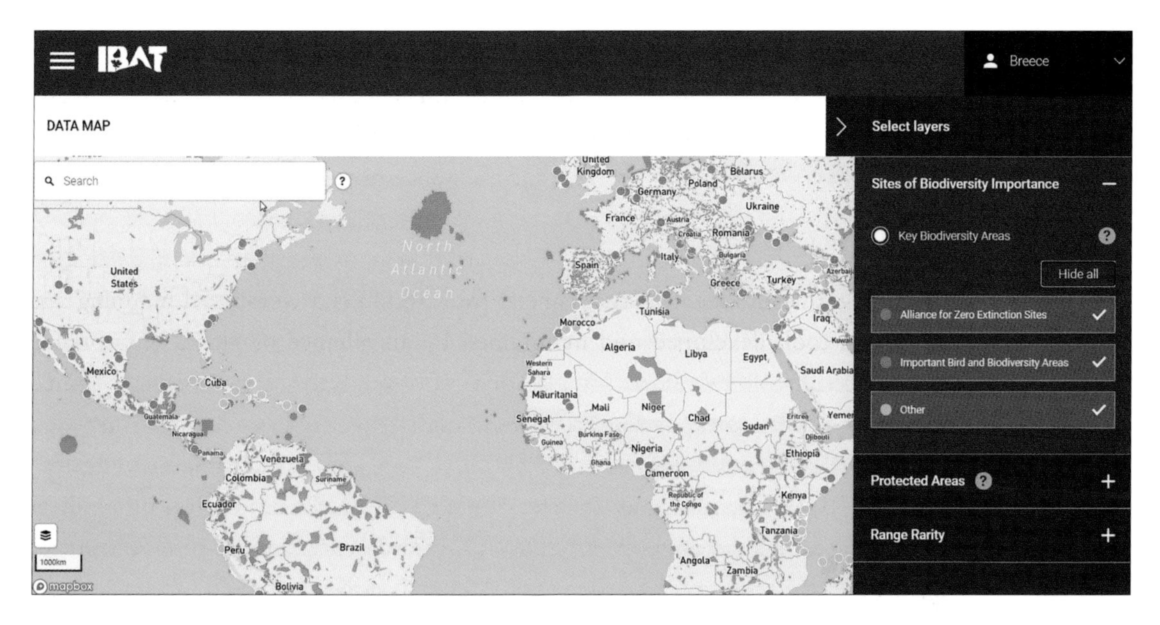

Figure 8.9. IBAT interactive online map showing biodiversity importance.

Agriculture & aquaculture ■ Biological resource use ■ Residential & commercial development
■ Invasive and other problematic species, genes & diseases ■ Pollution ■ Climate change & severe weather ■ Natural system modifications
■ Energy production & mining ■ Transportation & service corridors ■ Geological events ■ Human intrusions & disturbance ■ Other options

Amphibians
Birds
Fishes
Mammals
Reptiles
Arachnids
Corals
Crustaceans

Figure 8.10. Graph from a country report from the IBAT platform showing threats affecting different taxonomic groups assessed in a country. Only major classes of threats are shown.

Corporate data and AI initiatives

Companies are sharing their scientific, technology, and research expertise for conservation and environmental purposes through programs and initiatives. Only a few years ago, these types of resources were out of range for most conservation, environmental, and park organizations. The cost was exorbitant, and most organizations didn't have staff with the expertise to run big data projects. But the leadership and staff of many global companies understand the impact they can have on preserving the natural world. They also see the value and benefits that intact, highly functioning ecosystems and services have on their bottom line, as well as for the planet. These companies are the pioneers in developing and using big data and advanced technologies such as AI. Many are now supporting the conservation community through programs and initiatives.

AI and machine learning

AI is the use of advanced computer algorithms and computing infrastructure to solve complex problems by enabling data analysis at a scale exponentially beyond what people can accomplish with traditional computing methods. For example, imagine having to select and classify by hand hundreds of thousands of features such as trees from satellite images to perform a tree canopy analysis. Depending on the number of images and trees, this task could take weeks or even years to perform. AI can perform these same tasks in a fraction of the time. You can train a computer to detect trees, and then use GIS geoprocessing tools to classify or attach attributes to these trees. This machine learning process, in which users train computers to detect features, patterns, or relationships, is just one kind of AI. AI and machine learning can also return real-time results during disaster response and for wildlife protection and management.

Corporate partnership programs

Many companies offer AI programs that include grant funding and technical and infrastructure support for projects, programs, or initiatives. These companies provide engineering and data science assistance along with previously inaccessible data and cloud-based infrastructure. Microsoft's AI for Earth program supports hundreds of nonprofits such as NatureServe, to generate its high-resolution Map of Biodiversity (MoBi) maps, and OceanMind, to track illegal fishing. There are many more such initiatives, and I encourage you to explore what other companies are doing to support land protection efforts.

Arrangements such as these can be beneficial for both nonprofits and companies. Many staff at these companies want to help conservation organizations achieve their missions and make the world a better place. Conservation organizations benefit from having access to expertise and technology that would otherwise be out of reach. This type of synergy and collaboration is already advancing conservation and park projects and can be used to dramatically increase the pace and scale of land protection worldwide.

It can be challenging for organizations to find the time and resources for these relationships. Grants don't always cover staff time, and yet someone from the nonprofit must manage the project. It's worthwhile to consider the return on investment that a big data project can deliver to your organization and the conservation field. Often, an up-front investment in staff time to engage in these big data and technology projects can have a big return for your organization and for the conservation field at large. You can also consider partnering with a university to manage the project.

By combining GIS expertise with big data and advanced AI technologies and data science, we have the opportunity to reach our local to global conservation and park goals faster, and also to achieve the ambitious visions set out in the Half-Earth Project, the United Nations' Sustainable Development Goals, and the Thirty by Thirty Resolution to Save Nature.

Microsoft AI for Earth

AI for Earth is a five-year, $50 million Microsoft initiative using technology for sustainable solutions in agriculture, water, biodiversity, and climate change. Through the initiative, Microsoft's global elite researchers and engineers support many conservation groups by combining the power of the cloud, AI, and machine learning. The initiative has already created methods that greatly advance the pace and scale of conservation solutions. As we've seen, machine learning is used to produce high-resolution land-cover data in a fraction of the time it would take using traditional methods, and the higher resolutions help support better decision-making. These methods could also lead to processing other traditionally low-resolution datasets to help us make better land design and management decisions.

AI and machine learning are transforming how conservation groups can merge data from various sources such as satellite imagery, cameras, and drones to better understand species movements, behaviors, and stressors. The machine does the work for us, with algorithms that speed through various data and pinpoint the zebra herd or the power grid in minutes versus months. Check out the AI for Earth website for inspiring stories, to learn about partners and grant opportunities, and to access technical resources such as open-source tools, models, and APIs.

Amazon Web Services for nonprofits and NGOs

Amazon is supporting nonprofits through data, analytics, machine learning, and remote monitoring with the Internet of Things (IoT). The company's annual Amazon Web Services (AWS) Imagine Grant Program has funded many organizations working on human and environmental issues. The grants include a mix of direct support, AWS credits, marketing and storytelling support, training, and technical support. One grantee is using the AWS platform and machine learning to detect the practice of illegal bilge or oil dumping in oceans around the world, supporting countries with the tools they need to stop this source of devastating pollution in our oceans.

ArcGIS and AI

AI machine learning algorithms are applied through geoprocessing tools in ArcGIS Pro. The three areas of computation include classification, clustering, and prediction. The integration of machine learning into GIS means you can work a problem end to end in the GIS application. For example, you can use a machine learning tool to perform a cluster analysis on a large dataset and train the algorithm to cluster the features you are analyzing (e.g., biodiversity hot spots or water pollution sites). Then you can use GIS buffer and GeoEnrichment geoprocessing tools to analyze data around these clusters. The machine learning tools also provide statistical information and enable predictive analysis options. Esri is streamlining some of the AI and machine learning tools so that GIS experts can use them without having to be data science experts.

Maximizing the good of the mission

Datakind is a nonprofit organization that helps other nonprofits access the same resources that for-profits have for maximizing profits, except nonprofits are maximizing mission impacts. The organization connects organizations with expert data scientists to use big data, complex analytics, and advanced technology to advance the pace and scale of social and environmental impacts. Organizations can apply for pro bono support or join events that support short- or long-term engagements.

Putting it all together

There is no shortage of resources—from simple to complex, and from free to expensive—to support mapping and analytic needs. Much of the data that drives the free, online platforms highlighted in this chapter is available to download directly from the application sites or from ArcGIS Online or ArcGIS Living Atlas. This is important to note because you can directly download and integrate this data into your own maps and analyses by bringing it into ArcGIS Pro or into your own applications. To make it easy, organizations, agencies, or companies provide data in many formats for direct download or connection. When the problems you are trying to tackle are complex and require intensive data and analytics, it's worthwhile to explore online subscriptions or to apply for support from the companies that offer geospatial and AI options. These solutions can support your work and exponentially quicken the pace and impact of our collective land protection efforts.

CHAPTER 9

Using GIS to power and support organizational operations

Above, Ackerson Meadow in Yosemite National Park.

© Robb Hirsch, courtesy of The Trust for Public Land.

GIS can be a powerful system of support for organizational operations, although it's not typically thought of in this way. Unless people are trained in geography or GIS, it's difficult to truly understand how geospatial dynamics contribute to the underpinnings of the business and operations of an organization. For example, philanthropy teams have methods to find potential donors. They use data such as wealth, affinity, and willingness indicators to understand a donor's likelihood to give. But what about using GIS to show where donors with an affinity and ability to donate to conservation and parks live? They may live near a place your organization is raising money to protect, or they may be interested in a conservation fund-raising campaign for their neighborhood or community. These are examples of how map data can provide new insights and information to your philanthropy team and support new fund-raising tactics.

In this chapter, I'll provide an overview of ways to integrate GIS tools and geospatial thinking into organizational operations such as philanthropy, programs and initiatives, policy, advocacy, conservation finance, and marketing.

Philanthropy, fund-raising, and understanding your donor base

Celebrating and supporting your donors

Your organization is focused on important land protection efforts that attract the interests and passions of donors. Fund-raising teams need content and stories that inspire people to support the organization. GIS can help fund-raisers tell the stories of the organization through storytelling maps and apps such as ArcGIS StoryMaps.

Maps can show the impacts of your organization at whatever scale you work, from local to global. And if you work at multiple scales, maps show how your work nests and builds across geographies. Maps can show a project or a collection of projects and why they are important. Maps can focus on species and threats to their habitat. They may be organized around thematic programs, such as protecting agricultural and forest lands or working with tribes to repatriate ancestral lands. Maps can show an entire conservation corridor and the missing pieces within it. In all these examples, the goal of the map is to pique people's interest in your work and inspire them to support you.

A map to leave behind with donors can be a powerful fund-raising tool.

Some good examples of leave-behind maps

- Protected places or potential land protection projects (check confidentiality)
- Organization initiatives
- Map showing threats to the places you are trying to protect

Apps such as StoryMaps are also powerful tools for communicating with your board, donors, foundations, and volunteers. When people know a special place is protected, even if they never visit, it can stand as a beacon of hope. Maps can bring awareness of threats to places and issue a call to action to save special places or species that an individual may never personally see or experience firsthand. Personally, I love the National Audubon Society's classic StoryMaps story *Beating the Odds: A Year in the Life of a Piping Plover*. It's a moving and educational story about the lives of these birds and a call to action to protect their habitats (figure 9.1). You can find more examples from conservation and park organizations in the ArcGIS StoryMaps gallery.

Another way to engage people with your work is to create interactive mapping apps of the places your organization has protected. Many people like to know which organization or agency protected the places they like to go for a hike, to fish,

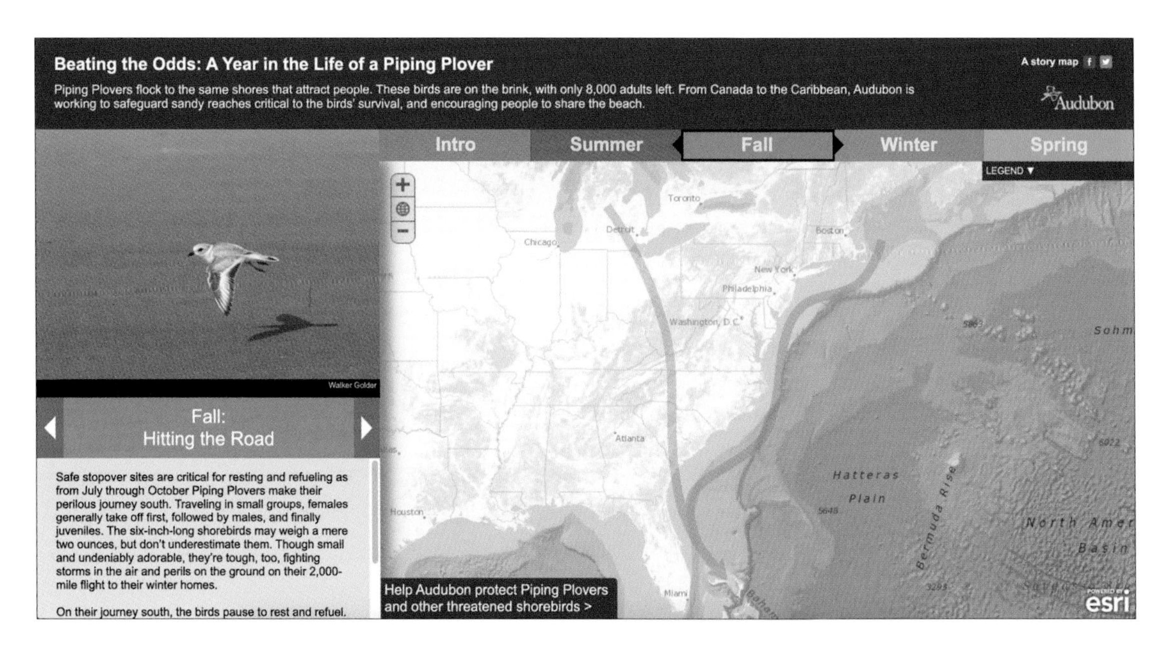

Figure 9.1. StoryMaps story following the migration of the piping plover from Canada to the Caribbean. This map is used to educate and engage communities about the importance of safeguarding and sharing the sandy beaches along the migration routes to save the species. There are only 8,000 adult piping plovers left, and the species is considered threatened and endangered.

The National Audubon Society.

or to walk their dog. People can also use these apps while traveling so they can visit the places you've protected. The Vermont Land Trust recreation story provides location information, pictures, and types of recreation activities for more than 150 sites it has protected across the state (figure 9.2).

Legacy maps and apps showcase the contributions of individuals or organizations. They can be focused on specific donors or volunteers to honor their contributions (but make sure to check with the donor for permission before sharing these maps publicly). One of my favorite storytelling maps showcases the legacy of Dr. Jane Goodall and the Jane Goodall Institute's Gombe Stream Research Center. The StoryMaps story is an overview of Dr. Goodall's early days as a researcher and includes interactive maps and information on her legacy in science, community engagement, and research, as well as current and future plans to protect the chimpanzees and their habitat while supporting the communities in the region to thrive and be a part of a sustainable future (figure 9.3).

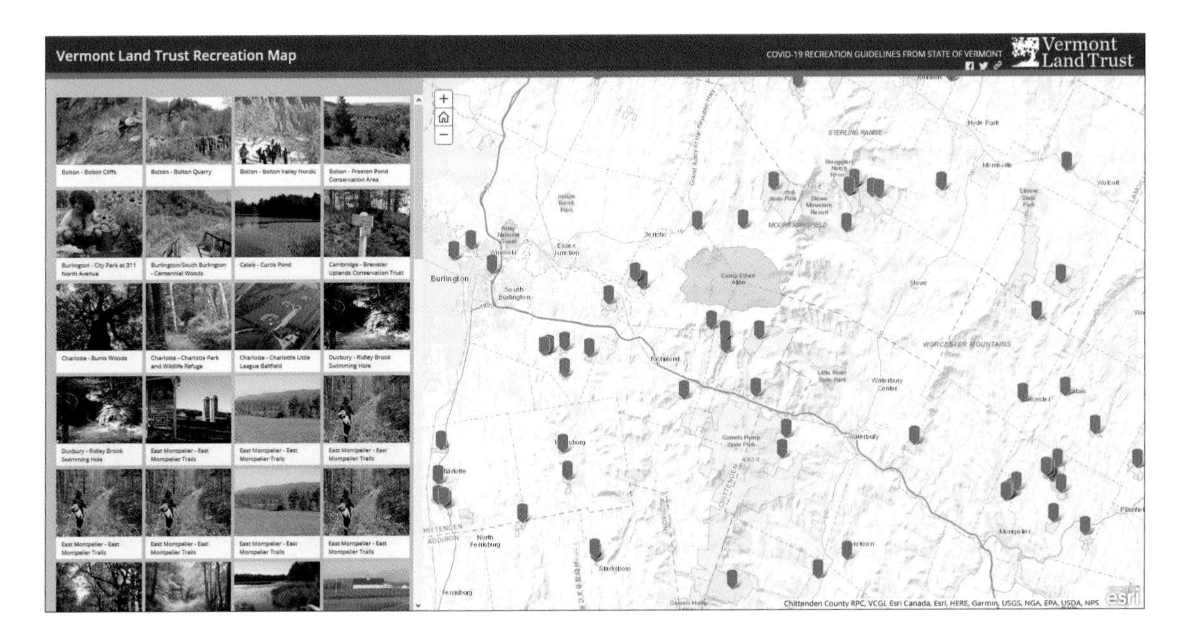

Figure 9.2. This Vermont Land Trust interactive StoryMaps story shows all the places the trust has protected, along with descriptions of each place and what visitors can do there. For example, at the Barre Town Forest, visitors can hike, ski, bicycle, and bird-watch.

Vermont Land Trust.

Figure 9.3. An example of a legacy StoryMaps story of Jane Goodall and Gombe.

Jane Goodall Institute.

Another legacy StoryMaps story example was created to honor the retirement of longtime Trust for Public Land (TPL) CEO Will Rogers. This story includes the history of Rogers' tenure with the organization, with stories and interactive maps showing land protection accomplishments (figure 9.4).

Another way to engage donors and volunteers is to create interactive maps to display on mobile devices during site visits. Interactive maps and apps complement paper maps, and they're easy to create out of the box through ArcGIS Online. You can add the data layer representing the property or sketch it on the map, add key data layers, and enable GPS tracking while on the property. That way, you know exactly where you are during your tour and can pair field observations with data to show the important aspects of the property.

Tip

If you are new to creating apps, look for the "Create an App" lesson at learn.arcgis.com. A lesson called "Hiking Red Rock Canyon" will walk you through the steps to create an educational app that provides information to hikers or walkers on how easy or strenuous the trails are.

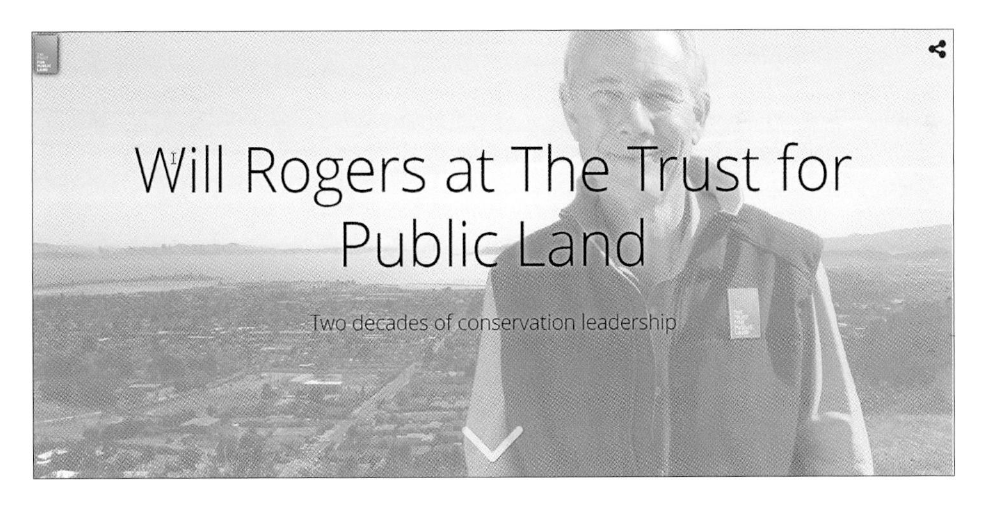

Figure 9.4. Will Rogers, president/CEO emeritus of TPL, in example of a legacy StoryMaps story.

Supporting your philanthropy team

Your fund-raising teams are working hard to find new connections and potential donors for your organization. Some organizations have sophisticated donor management software that helps them assess, track, and manage existing and potential donor relationships. Most of these applications either have no mapping component or one that's limited in what it can track and analyze. Some software packages offer profiles of communities at different geographic scales with information on spending preferences, nonprofit giving history, and likelihood of giving to your organization. This information helps philanthropy teams target outreach activities. But these software packages are so expensive that most nonprofits can't afford the licenses. Even large organizations still end up doing a lot of hands-on work and online research to fill in important information missing from the donor management software and to identify which communities to target with outreach.

Finding a new way that GIS can help

Early in my career at TPL, I discovered a way that GIS could help philanthropy teams. Staff had spent weeks transcribing demographic information they'd found online onto paper maps to identify zip codes and addresses for mailing an event flier. They didn't know that I could create this map for them in minutes using GIS. I realized then that I needed to do more to educate teams in the organization about GIS and what it could do for them, and to learn more about their work to understand where I could suggest GIS-based solutions.

Geospatial software companies have also developed tools to help companies and organizations understand their constituents. Esri® Tapestry Segmentation groups neighborhoods based on lifestyle choices and consumer spending habits. It pulls information on restaurant preferences, purchases, dog ownership, car ownership, vacation destinations, and much more. The product classifies neighborhoods into 67 segments, including Laptops to Lattes, Rustic Outposts, Ethnic Enclaves, and Metro Renters. Search for the Tapestry Segmentation poster that includes all the summary groups along with a breakdown of data for each group. This database can be used to create a Tapestry Segment profile or a combination of segments that fit the profile of conservation and park supporters (figure 9.5).

You can overlay tapestry segments with completed land protection projects to identify the profiles that match where money came from to support your projects. This information can help you better understand that donor base. You can then overlay potential or active projects and apply that profile to help direct fund-raising campaigns or targeted mailings.

Look for the StoryMaps story *How to Use and Interpret Esri's Tapestry Data* for an overview of the Tapestry market segments, LifeMode summary groups, and urbanization groups. It includes examples of how to use and interpret the data for siting a new business. You can substitute a park or open space for the business and see how the data can provide a more comprehensive understanding of who will be served by new protected lands, who is already served, and who could become supporters of your organization.

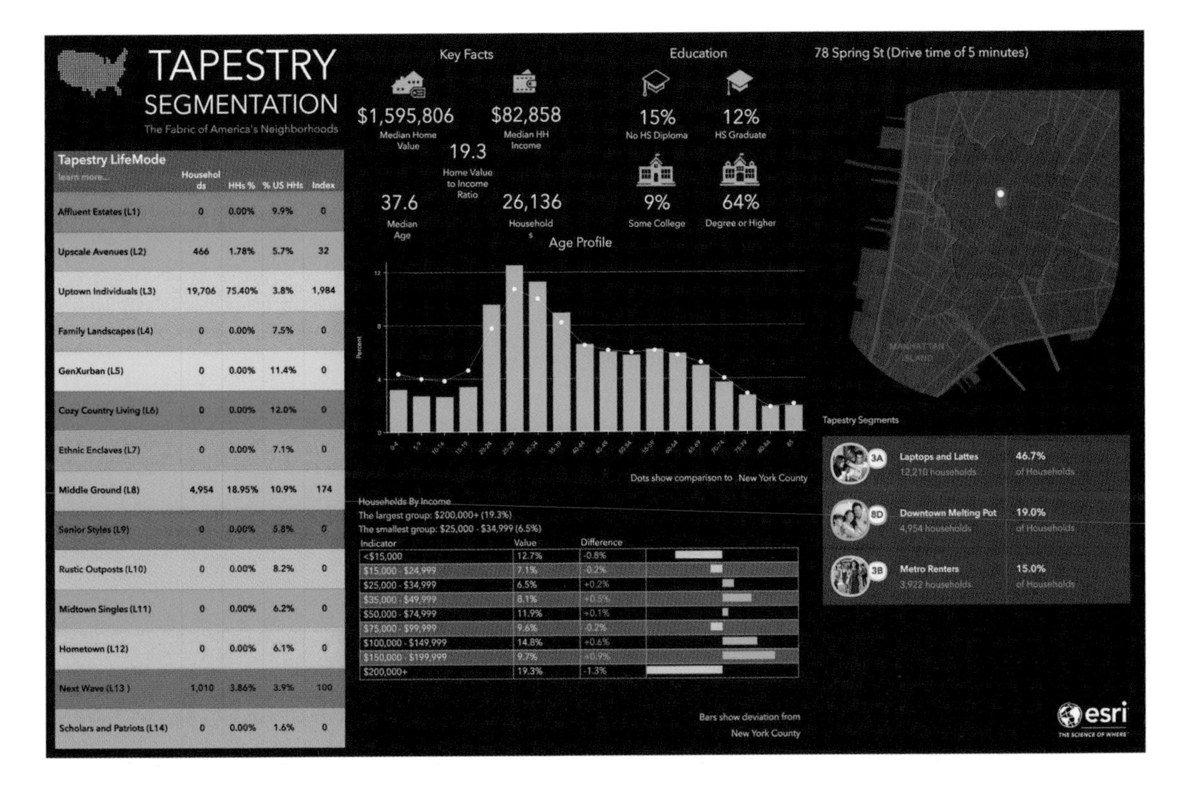

Figure 9.5. Example of a Tapestry Segmentation profile in lower Manhattan. This area is characterized by 46.7% of the households falling into the Laptops and Lattes segment. The report includes information such as median age, education, home values, and more.

Strategic conservation plans and maps give supporters confidence

In chapter 6, we discussed using GIS to drive organizational strategy, but it's worth highlighting again here. Supporters and partners want to know that their money is being used to save the places with the highest conservation values. They also want to know that they are supporting land protection programs and operations that are effective. Your maps help support these stories and bring financial resources to the organization. For example, your ability to prioritize protecting one place over another could mean the difference between getting donations, grants, or funds for your projects and having those resources go to another organization. If your data demonstrates how one of your programs affects community health or is saving a species, that could help you close a foundation grant.

Supporting organization themes, programs, and initiatives

Using GIS to scope and define programs

Every land trust and conservation organization has a way to structure activities and resources. Some call them focal areas, and some call them themes, programs, or initiatives. For the purposes of this chapter, I'll refer to these organizational categories as *programs*. Programs are based on topics such as wildlife habitat, climate resilience, social equity, or connectivity. GIS provides guidance on where an organization should focus, based on their program.

Examples of conservation organization programs

- Conservation International—secure healthy ecosystems, ensuring critical areas remain intact
- American Forests—urban forestry, ensuring tree equity in American cities
- Surfrider Foundation—coastal preservation, protecting shorelines from threats such as dredging, seawalls, and more
- Center for Large Landscape Conservation—Corridors and Crossings, creating safe passage for people and wildlife
- Great Land Trust in Southcentral Alaska—habitat, ensuring salmon habitat conservation
- African Parks—saving wildlife, ensuring wildlife protection and restoration

Creating maps and products for programs can be as simple as identifying the focal area of the program and including data related to the theme of the program. GIS products can also be robust and comprehensive, such as a data-driven programmatic plan. Program-based GIS tools can also turn into innovative solutions for broader use. Many useful tools and platforms have been created out of organizational programs such as Global Forest Watch, covered in chapter 5.

Another example of an innovation that started as a program is the Ocean Health Index (OHI), a science-based decision support platform to protect ocean ecosystems and the humans that depend on them. The index measures ocean health on the basis of key benefits the ocean provides people and how communities that depend on oceans for livelihood and quality of life are protecting and sustaining them. OHI was developed by partners including Conservation International and the National Center for Ecological Analysis and Synthesis.

Using GIS to map your programmatic areas helps you understand where your organization is working in relation to others. This can help avoid overlap and duplication, align resources, and identify partnership opportunities to maximize conservation impacts. For example, the conservation easement work of a land trust is a key contributor to the land acquisition work being done by other organizations and agencies. Incorporating the programmatic focus of the work of your organization into the work being done by others in a region or area focuses and aligns resources and maximizes conservation impacts, such as protecting the right lands for water quality or wildlife corridors. This type of coordination and collaboration demonstrates to funders that your work will protect the places of greatest need.

Using GIS to measure program impact

The field of conservation biology learned early on how to integrate GIS into planning and decision-making. It has also developed a robust system of methods to evaluate and monitor habitat to assess the progress being made in protecting species. These methods influence how lands are managed and monitored to assess conservation successes and identify the need for new approaches. The proliferation of high-resolution data, beyond traditional habitat and species data, broadens the use of data for measurement, evaluation, and real-time management and operations. If you have a forest protection program, you can choose from many high-resolution tree cover and canopy datasets, scaled from local to global, to integrate into your analytical approach. The same goes for ocean protection, urban parks, and climate work. GIS is an end-to-end support platform for mapping and analyzing in planning, evaluating, measuring, and managing programs.

Policy and advocacy

Land protection is only the first step in conserving and preserving the places we love. We must follow up with policies that stop, reverse, or adapt human uses or influences that cause degradation of habitat, species loss, overfishing, overuse, and other factors that diminish ecosystem values (figure 9.6).

Policy

Conservation policy includes mechanisms, principles, or actions used as a basis to make decisions for the protection and restoration of human and natural ecosystems. Policy can be complicated. We hear all too often about challenges that can dismantle or weaken the rules, laws, and regulations that are in place to protect places and species. Unless you can decipher technical information and legal language, it's hard to understand what these challenges mean. That's why communicating science, data, and policy to nonexperts is a key part of advocacy to ensure that land protection

Figure 9.6. "Protect Our Public Lands" sign at the Rally for Bears Ears and Public Land in Salt Lake City, Utah, on December 2, 2017.

rules and regulations are upheld and strengthened. Telling the right stories about proposed and real changes that threaten conservation and park policy can galvanize communities and politicians to stop those changes or create new and more supportive policies.

It's important to get the right teams and partners together to use data and maps to understand the implications of new or revised policy. We can't do this work alone. It takes all of us working together to advocate for the best policies that support our missions. GIS products can help unite partners with various backgrounds and expertise to agree on the potential impacts of change or the need for new policies.

Figure 9.7 shows how GIS supports advocacy and policy efforts and the interplay between the two. GIS mapping and analysis is used to show on maps where to prioritize funding for policy investments. For example, maps can show where low-income communities experience higher rates of asthma and pinpoint neighborhoods where policy investments in green infrastructure and trees are needed. Next, GIS analysis can inform where funding is being directed that is not solving the issues

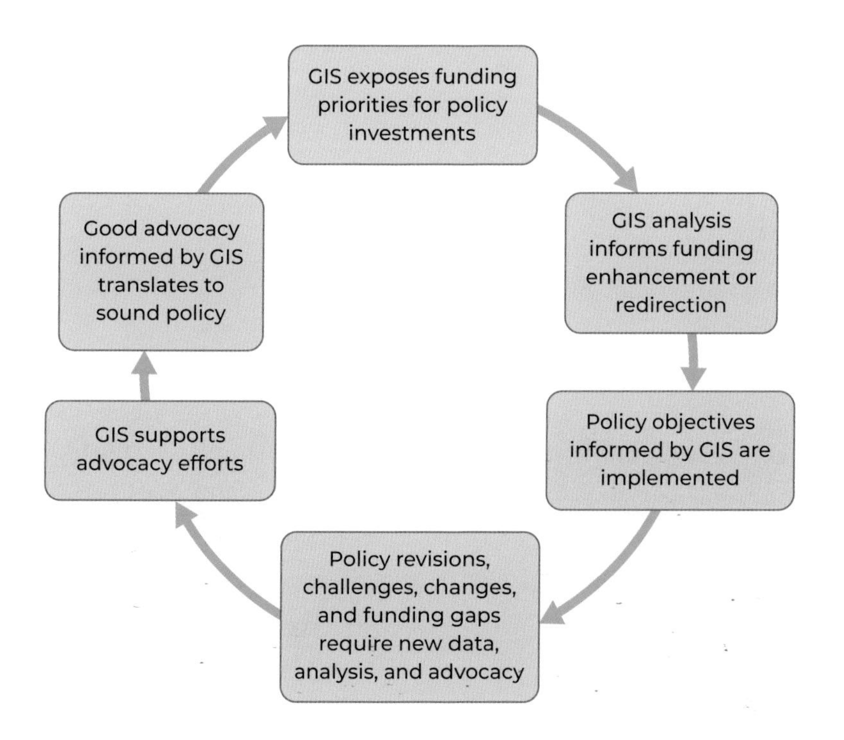

Figure 9.7. How GIS supports advocacy, policy, and funding.

David Weinstein and Breece Robertson.

at hand or where there is no direct funding where needed to solve the problem. For example, maps can show where funding is being directed to protect land. The maps can be used to overlay data such as Map of Biodiversity imperiled species to determine whether land protection funding is being directed to the places that could support animals and plants in decline.

By informing funding priorities with GIS, policies can be implemented that better match the intended and needed outcomes. GIS can and should be used to inform policy revisions by updating data and maps over time so funding can be redirected or created to meet the changing dynamics of communities—environmental, social, and economic. Changing policy and redirecting or creating new funding usually requires advocacy to educate and inspire people about the cause or issue, such as improving equitable access to park and public lands or keeping public lands in public ownership. Maps, dashboards, and StoryMaps stories are important tools that advocacy organizations use to support changing, defending, or revising policy when needed.

Strong and consistent conservation policies from the local to the global scale are critical to ensure that protected lands are managed and used in sustainable and responsible ways.

Figure 9.8. Bears Ears Buttes from Snow Flat.

Josh Ewing, Friends of Cedar Mesa.

Conservation organizations focus on different areas of policy. For example, Conservation International focuses on policy in the areas of ocean health, deforestation-free commodity supply chains, and combating wildlife trafficking, among others. The National Resources Defense Council fights oil and gas drilling in wild places. The Indigenous Environmental Network focuses its policy efforts on fighting fracking, mining, and oil and tar sands drilling to protect lands. Friends of Cedar Mesa, in southwest Utah, works for greater protection of Cedar Mesa and the Bears Ears National Monument by supporting local to national policy efforts (figure 9.8).

One example of how organizations use maps to support policy and advocacy is the Michigan Land Trust interactive app. Ducks Unlimited created an interactive web map to help the land protection community in Michigan visualize trends over time and to support planning efforts. The app uses freely available data from the National Conservation Easement Database and the US Geological Survey (USGS) Protected Areas Database of the United States. You can visualize trends and accomplishments by county, US House district, US Senate district, and conservancy service areas. These maps support policy and advocacy efforts by land trusts in the state (figures 9.9 and 9.10).

Advocacy

Many people confuse advocacy with lobbying. Advocacy is the active support of a cause, whereas lobbying involves attempting to influence legislation or politicians. Nonprofits in the United States can lobby for legislation with some limitations. Look online for resources that explain laws and definitions, and, if necessary, consult legal counsel on the issue of lobbying.

Advocacy is focused action to raise awareness and move people to act on behalf of your mission or campaign. Every organization can advocate, and I would argue that all conservation organizations should. Protecting the lands needed to slow climate change and promote sustainability is a big job. Advocacy done right educates and raises awareness about an issue, influences decision-makers, and can create real change. Following are a few examples of organizations with strong advocacy programs.

The Outdoor Alliance (OA) is a coalition focused on the protection and management of public lands for "human-powered" uses such as biking, skiing and hiking. The alliance's GIS Lab creates map products to support advocacy and policy. Look for its StoryMaps story *Turning Data into Advocacy* for examples of how the

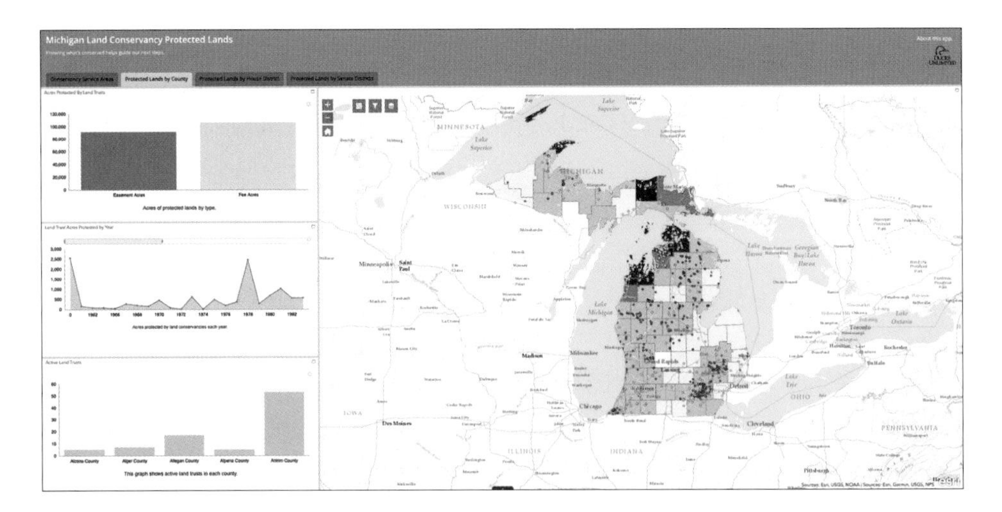

Figure 9.9. Michigan Land Trust interactive map shows data about land protection by county. The graphs from top to bottom show acres of protected lands displayed by easement (dark blue) or fee lands (light blue), acres conserved by land conservancies annually, and number of active land trusts in each county.

Ducks Unlimited.

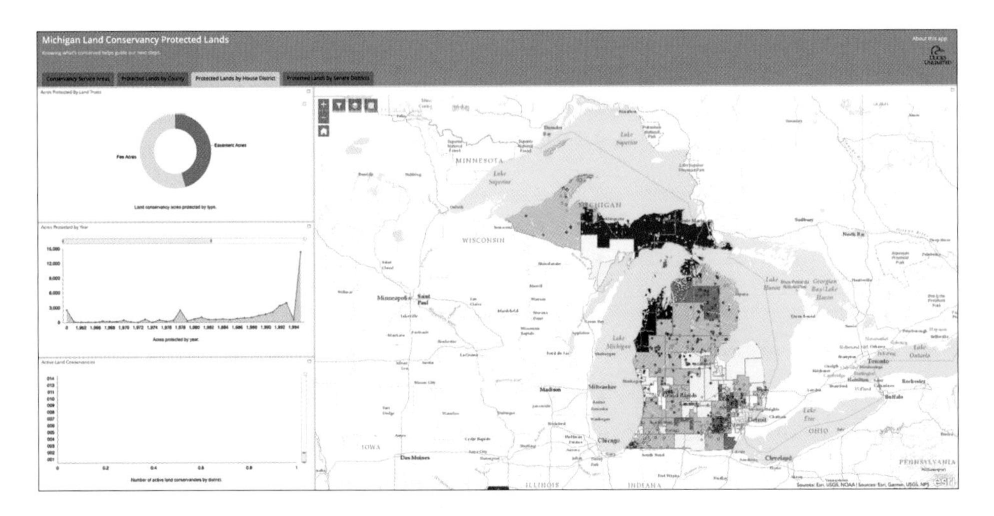

Figure 9.10. Michigan Land Trust interactive map shows land protection by House district. The graphs from top to bottom show acres of protected lands displayed by easement (dark blue) or fee lands (light blue), acres conserved by land conservancies annually and number of active land trusts in each House district.

Ducks Unlimited.

coalition uses GIS to support important issues such as protecting national monuments and analyzing the economic impacts of outdoor recreation. Also, look for its page with informative GIS maps and apps on advocacy, forest planning, proposed legislation, and public lands for examples on how they approach using GIS for policy and advocacy (figure 9.11).

Figure 9.11. Outdoor Alliance's "maps and apps" web page is a resource for the outdoor recreation community that contains web mapping apps and maps related to OA's campaigns and projects.

Outdoor Alliance.

The Grand Canyon Trust uses GIS, science, and data to protect the Grand Canyon and the Colorado Plateau. Its mission is to protect the air, water, wildlife, and habitats of the canyons, forests, and mesas while supporting the rights of its Native peoples. The trust uses maps to support advocacy and litigation to prevent land-use changes on the plateau that would harm the land, the wildlife, and the people. Figure 9.12 is an example of using data to educate people about threats by showing the extent of energy development near the Grand Canyon.

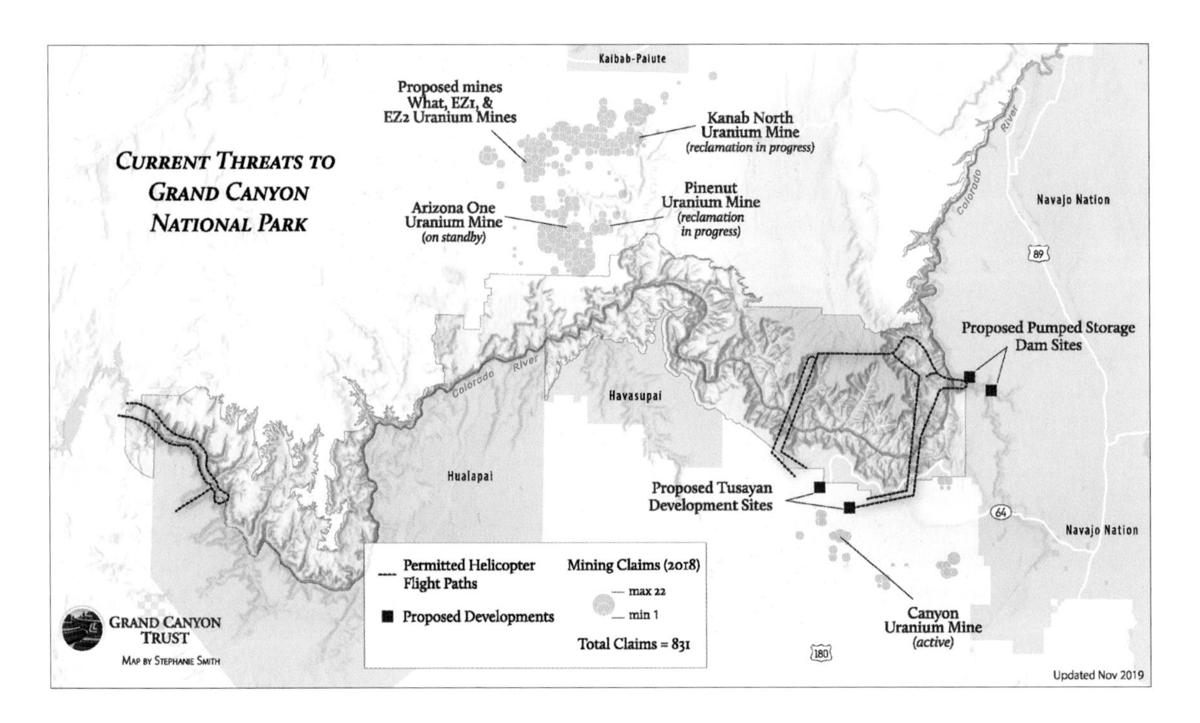

Figure 9.12. Grand Canyon National Park may rank as one of the Seven Natural Wonders of the World, but national park status doesn't give the canyon the complete protection it deserves. Hundreds of daily helicopter tours over the canyon turn the canyon calm into a buzzing hum. A legacy of uranium mining and development proposals, from tourist resorts to dams, seek to profit from the natural resources on lands surrounding the park, leaving seeps, springs, and precious waterways to chance while also threatening cultural resources important to many Native peoples.

Stephanie Smith, Grand Canyon Trust.

How GIS can help advocacy

Early in my career with TPL, I realized that staff working on policy and advocacy were printing 1:24,000 USGS maps, tracing the boundaries of specific properties on the maps, and using markers to color potential conservation properties on the basis of ownership. They would then include these maps in funding proposals or take them to meetings with congressional members or agency staff. They were developing hundreds of these maps a year during the crunch time when the legislature was in session.

These are tasks that GIS can accomplish much faster, with greater polish and accuracy. Many organizations are still doing this type of work with markers and printed maps. Our organizations could be much more efficient if we all identified tasks that could be automated, and then created a GIS solution, leaving policy and advocacy staff with more time to think and act strategically.

Park staff could engage more deeply with the community instead of searching for demographic data and trying to guess how many people their parks are serving. GIS supports efficiency, effectiveness, and the right use of resources for the right tasks.

Conservation finance

Conservation finance is the practice of creating and managing funding to support land, water, and natural resource conservation. Globally, around $50 billion has been generated through conservation finance, but that falls far short of the estimated $300 billion to $400 billion needed to meet critical conservation milestones (Martin 2015). It's worthwhile to learn more about conservation finance—the mechanisms, challenges, and innovative approaches that organizations are undertaking to create funding to preserve places before it's too late. I recommend resources from the Lincoln Institute for Land Policy, Conservation Finance Network, International Union for Conservation of Nature, and Conservation Finance Alliance. Some conservation organizations that focus on conservation finance are TPL, The Conservation Fund, and The Nature Conservancy.

Tried-and-true guide for conservation finance

The World Wildlife Fund produced a report on global conservation finance in 2009. Although it's older, the report contains in-depth information about conservation finance mechanisms that hold true today. If you are new to conservation finance, this report provides a comprehensive overview of mechanisms such as carbon financing, watershed services, fishing, tourism and recreation, fees from real estate development, and much more, including case studies with successes, challenges, and innovations.

A conservation finance campaign is a multistep process. A lot of time goes into research, such as identifying funding needs; understanding place-specific laws; understanding how other communities of similar makeup have approached, succeeded, or failed at conservation finance campaigns; and understanding public opinion. Feasibility assessments and polls test the public's support of the measure and inform communications. Then good ballot language must be developed to ensure that the intent resonates and is clear, concise, and understandable by the public. Almost all these elements are influenced by geospatial dynamics and can be mapped and analyzed. For example, we can now identify which communities have a high probability of supporting park and open space bonds by mapping income data, voting histories, and consumer preferences. By integrating GIS into the feasibility and research process, experts have more information that drives better conservation finance outcomes.

Uses of GIS for conservation finance

- GIS maps can inform decisions about directing or redirecting funding to preserve the places that will meet specific policy goals.
- GIS can help policy enforcers identify where conservation funds are being diverted into other interests when they shouldn't be.
- Nonprofits and agencies can use GIS to determine how to direct voter-approved funding for parks and conservation.
- Conservation finance campaign leaders can use GIS to make the case for more federal or voter-approved funding for specific locations or causes.

A track record for conservation finance

TPL published *The Conservation Finance Handbook* in 2004, and it is still considered a premier resource for a step-by-step approach to designing a measure and conducting a campaign. To date, TPL and partners have worked on more than 700 ballot measures, with 576 passing—an 82 percent success rate—to create $79.98 billion in voter-approved public funding for parks, land conservation, and restoration.

Use case 1: Funding strategic conservation priorities

In this case, you have a strategic park plan that identifies places to protect. The community supports the plan. Now you need the money for land protection. Existing sources of funding aren't sufficient to cover new land protection purchases, but the public is eager for more outdoor places to play, relax, and recreate. Your organization has identified a park ballot measure as the funding mechanism to fill the financial gap needed to implement the plan.

First, use TPL's Conservation Almanac to view conservation spending trends and programs in your state. This free, online mapping app and database tracks public conservation spending, acres conserved, sources of funds, and more, for federal, state, and local levels. You can view a chart for your state showing conservation dollars contributed annually by state, local, or federal sources to help you understand where most conservation funding is coming from (figure 9.13).

You can also explore protected lands on the interactive map to get project-by-project information on funding sources, year, and amount spent. You can use the almanac to identify matching funding sources for a ballot measure or highlight the need for future new or reinvestment opportunities. Look for TPL's Conservation Almanac StoryMaps story for more on this online tool (figure 9.14).

Next, use TPL's LandVote database to see which ballot measures in your state have passed or failed. You can also search nationwide for communities with similar characteristics to yours that have passed ballot measures. This database dates to 1992 and is updated annually (figure 9.15).

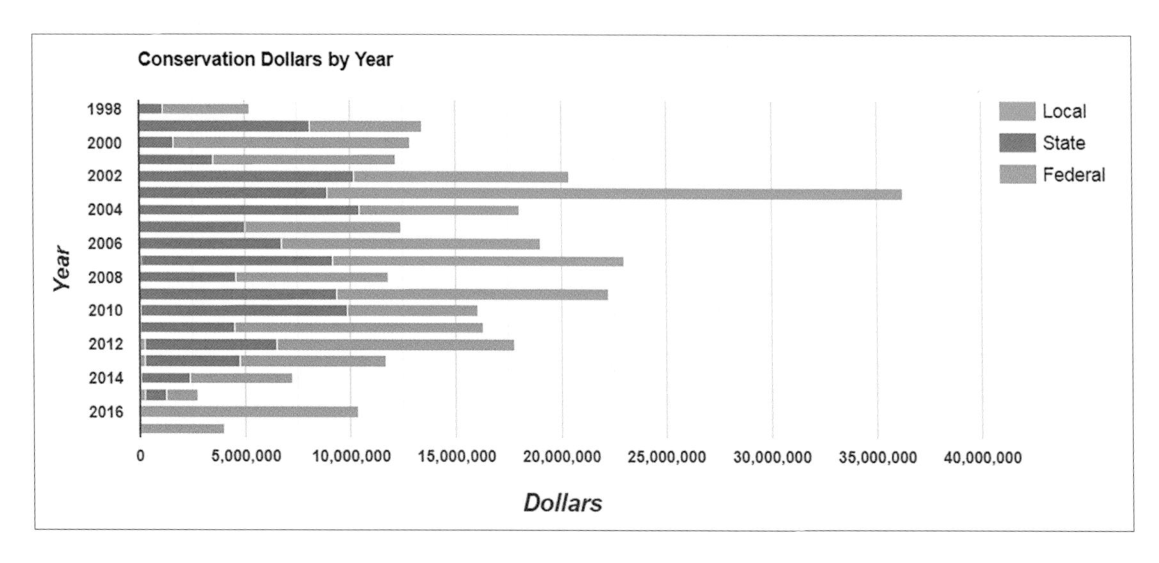

Figure 9.13. Conservation Dollars by Year for Maine. Conservation Almanac, 2019, www.conservationalmanac.org.

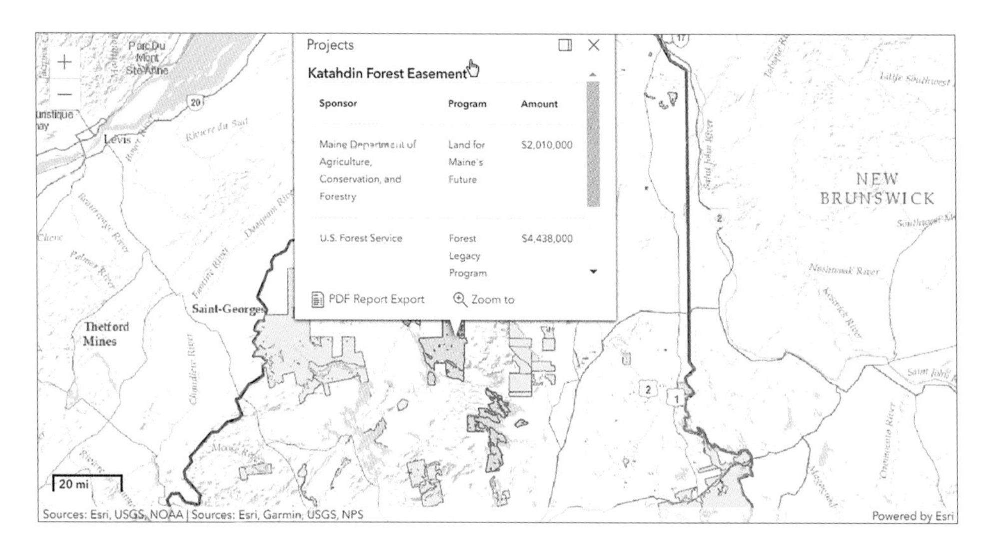

Figure 9.14. Using the Conservation Almanac web app, you can click on a protected place and see the funding information. In this example, the pop-up shows the funders or sponsors (i.e., US Forest Service), the funding program (i.e., Forest Legacy Program), and the funding amount from each funder or sponsor for the Katahdin Forest Easement.

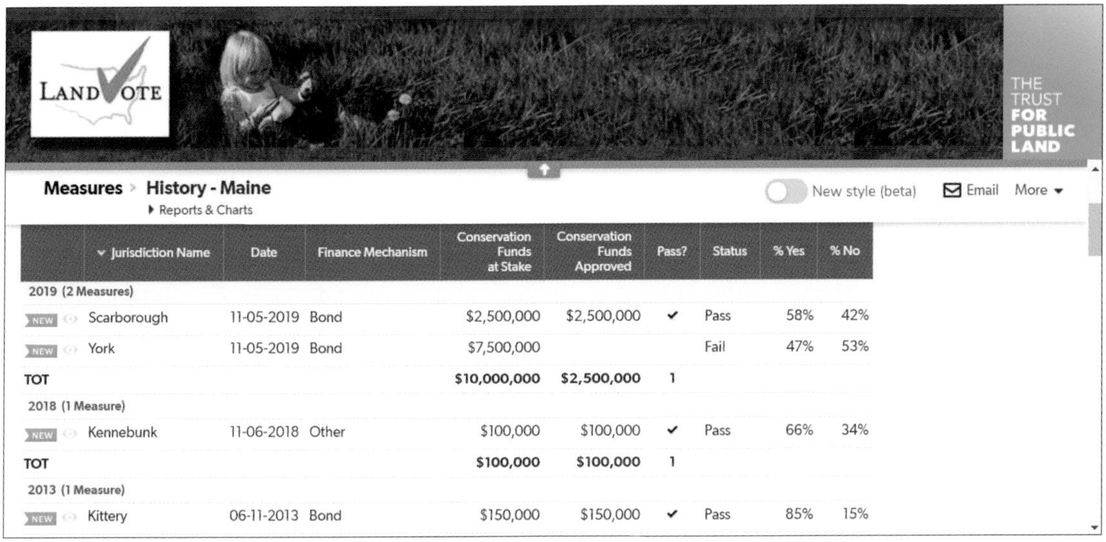

LANDVOTE

THE TRUST FOR PUBLIC LAND

Measures > **History - Maine**

Reports & Charts

New style (beta) Email More

Jurisdiction Name	Date	Finance Mechanism	Conservation Funds at Stake	Conservation Funds Approved	Pass?	Status	% Yes	% No
2019 (2 Measures)								
NEW Scarborough	11-05-2019	Bond	$2,500,000	$2,500,000	✔	Pass	58%	42%
NEW York	11-05-2019	Bond	$7,500,000			Fail	47%	53%
TOT			**$10,000,000**	**$2,500,000**	1			
2018 (1 Measure)								
NEW Kennebunk	11-06-2018	Other	$100,000	$100,000	✔	Pass	66%	34%
TOT			**$100,000**	**$100,000**	1			
2013 (1 Measure)								
NEW Kittery	06-11-2013	Bond	$150,000	$150,000	✔	Pass	85%	15%

Figure 9.15. LandVote database shows conservation ballot measures that have passed or failed in the United States.

Now that you have used two free online resources for data on conservation spending sources and past ballot measures, you can help conservation finance researchers by providing maps and analytics on demographic, socioeconomic, spending, consumer preference, environmental, and other data, including threats, that can support the research efforts.

Use case 2: Feasibility study

In this case, your community does not have a strategic plan that identifies where lands should be protected. Your community is experiencing a dramatic increase in human and energy development, which threaten the most pristine, intact lands. A group of citizens wants to protect these lands and wants to explore the feasibility of a park ballot measure. To succeed, your organization must engage the broader community and tell the story about what will be lost, given the rapid changes, if funding isn't available to protect lands.

Maps showcasing the pristine places and threats often spur people into action. Esri® Green Infrastructure Assets data layers through ArcGIS Living Atlas of the World and ArcGIS Online can help you create maps that show what is at stake if this measure doesn't pass. You can also create interactive web maps or StoryMaps stories in ArcGIS Online so the information can be shared widely and integrated into news stories and outlets. The Esri Green Infrastructure Initiative, TPL Greenprint method, and other conservation planning processes will guide you if you need to create a multibenefit park and open space plan to support the effort. Esri's green infrastructure apps (figure 9.16) provide advanced tools for landscape analysis, including the ability to select, prioritize, and investigate intact habitat cores and perform your own landscape analysis by prioritizing and filtering data that represents how humans have modified the landscape, where to prioritize land protection to support biodiversity, and where there is a variety of landforms that support plant and animal species, among many other types.

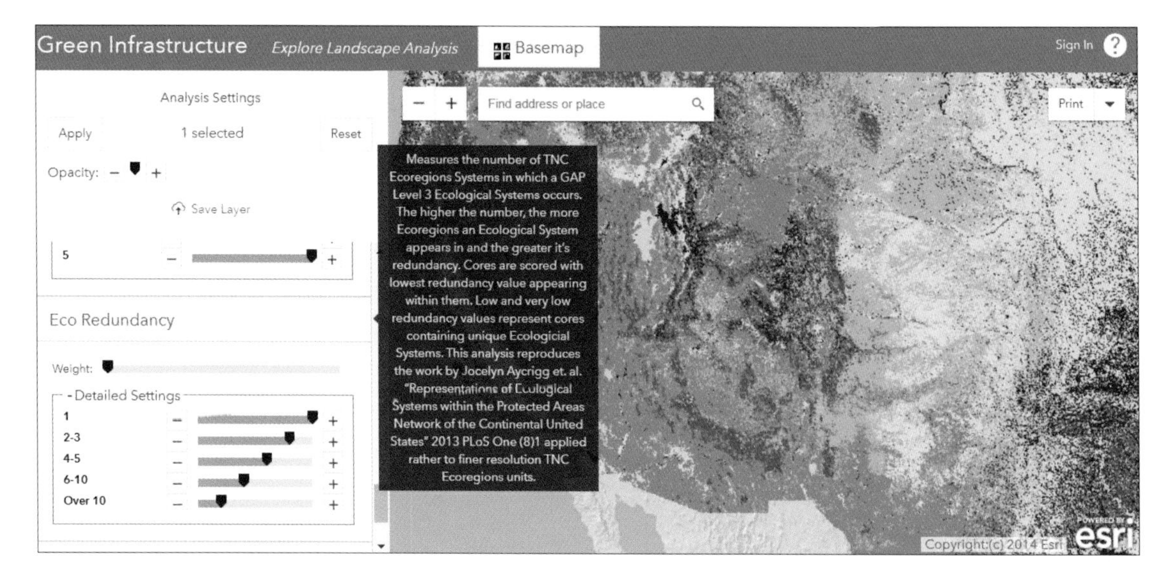

Figure 9.16. The Esri Green Infrastructure Explore Landscape Analysis app shows the filtering and prioritization tools for the Green Infrastructure datasets. In this image, the Eco Redundancy layer is displayed and symbolized using a green color ramp. The definition of the layer is displayed in the gray box, and the slider bars allow the user to weight either the layers or the subitems for those layers.

Future opportunities

Market-based funding mechanisms show promising signs as a complement to traditional conservation finance. These mechanisms tend to be more stable and dependable, enabling conservation organizations to be more proactive and strategic, which is needed to meet the pace and scale required to save the places essential to a sustainable future. Private-sector investors are increasingly interested in supporting efforts that will provide a financial return in addition to benefiting social, economic, and environmental causes. This is referred to as "impact investing." A study by NatureVest and Eko Asset Management Partners showed rapid growth since 2004 in these types of market-based conservation impact investments. This is a promising area to watch and engage with to diversify and increase conservation funding (Martin 2015).

One of many challenges facing market-based conservation investments is the lack of widely accepted standards for measuring impacts (Davies et al. 2019). This is certainly an area in which GIS can be scaled up. An exciting and promising effort by the United Nations (UN) Sustainable Development Solutions Network, Esri, and the National Geographic Society is a portal called SDGs Today (figure 9.17) for data on the UN's sustainable development goals (SDGs). This platform supports the 2030 Agenda for Sustainable Development blueprint, which was adopted in 2015 by UN member states. The blueprint is a guide that supports actions to achieve the goal of peace and prosperity for people and the planet. The portal tracks real-time and regularly updated data to measure progress on meeting the SDGs. By tracking and visualizing data, the portal enhances our understanding of what we need to do to increase the pace and scale of solutions needed to meet these goals.

Marketing and communications

All effective organizations and agencies market or communicate their work. Making GIS a key part of the marketing and communications strategy will help your organization reach more people, tell a better story, and engage supporters and advocates. It's important to work with your marketing and communications team to maximize storytelling and reach a broader audience with GIS maps and products. Use organizational branding to create map styles and templates. Your GIS products should adhere to your organization's brand standards, so users immediately associate the product with your organization and your mission.

Figure 9.17. In partnership with Esri and the National Geographic Society, SDGs Today: The Global Hub for Real-Time SDG Data is a platform developed by the Sustainable Development Solutions Network. This open-access data platform provides a much needed virtual space where key stakeholders from around the world can access and engage with timely data (updated annually or in more frequent intervals) on the SDGs and learn how to use the data effectively to push forward Agenda 2030 for Sustainable Development.

Sustainable Development Solutions Network.

Marketing teams can blend maps and data into inspiring stories. It's helpful to get support to transform data and technical subjects into stories that a non-GIS expert can understand and relate to. Sometimes we are just too deep in the data weeds to grasp the big picture and how it connects to what else is going on in the organization and the world at large. A good relationship with the marketing team can go a long way for storytelling with data.

GIS web technologies such as interactive maps and maps that tell stories can help organizations expand their impact, presence, and reach. But blending GIS web products with marketing technologies requires strategy, collaboration, and leadership buy-in. Many organizations have a digital road map that includes all the ways the organization uses online technologies, products, and social media. Unless you have a relationship with the marketing team, GIS may not be a part of that road map. Meet regularly with the marketing team to ensure that GIS is part of this digital road map, with a focus on integrating GIS into your organization's social media strategy.

In this chapter, I've covered just a few examples of how GIS can support organizational operations more broadly. Whether your organization is small or large, GIS brings efficiencies to internal workflows or can take a program or campaign to the next level. It's worthwhile to talk to your colleagues and discover where GIS might help. It can be in the form of supporting better storytelling or using advanced data and analytics to understand what drives people to support saving places. Better alignment and integration within organizations means bigger impacts, a broader reach, and more effective coordination of land protection efforts. The urgency and pace that is needed to save our world means it is essential that we are creative and use all our tools and technologies. Using GIS to support internal operations optimizes resources with the goal of more land saved.

References

Davies, R., H. Engel, J. Käppeli, and T. Wintner. 2019. "Taking Conservation Finance to Scale." McKinsey & Company. Accessed January 3, 2021. https://www.mckinsey.com/business-functions/sustainability/our-insights/taking-conservation-finance-to-scale.

Martin, Chris. 2015. "Conservation Finance 101." Conservation Finance Network. Accessed January 3, 2021. https://www.conservationfinancenetwork.org/conservation-finance-101.

CHAPTER 10

The road ahead

Above, aerial view of Honouliuli Preserve, Waianae Mountains, in Oahu, Hawaii, in 2002.

©Phil Spalding III, The Nature Conservancy, courtesy of The Trust for Public Land.

I hope you are inspired to either get started with GIS or step up your use, and that you understand why GIS is a crucial tool for protecting more land faster and managing those lands more effectively. I've covered using GIS to address conservation and park issues, including storytelling, applied analytics, data science, measuring progress, and community engagement. I've only scratched the surface of the ways that people are using GIS for land protection across the globe, so I hope you will reach out to organizations and park agencies in your area.

The field of GIS evolves and changes so rapidly that what was considered innovative yesterday is mainstream today. Many of the tools and approaches highlighted in this book required deep expertise to use and were time-consuming and expensive to develop. But today, a lot of the data, methods, and applications are available to everyone, often for free online or through subscription services.

App development is one example of this trajectory. Just a few years ago, most organizations had to hire a consulting firm and spend hundreds of thousands of dollars to develop apps. Today, you can create simple apps quickly with no programming experience.

Advances in computing power and data storage and the move to cloud-native and enterprise systems are broadening the possibilities for GIS innovations. Innovative ideas become reality faster, more efficiently, and at lower cost. For example, the Chesapeake Conservancy's Conservation Innovation Center (CIC) partnered with Microsoft to combine machine learning in the Azure cloud with Esri technology to create one-meter, high-resolution land-cover data for the entire Chesapeake Bay watershed—a vast improvement over the previous 30-meter scale. It provides much greater precision in water quality planning and conservation prioritization. Technology helped CIC create a dataset within months that would have taken years to develop using traditional GIS methods.

Advances in technology and GIS are supporting stewardship and fieldwork, from local parks to large open spaces. Field apps used to be hard to set up and clunky to use, didn't integrate data easily when you got back to the office, and didn't work unless you had a cell phone signal or Wi-Fi. Moreover, most didn't integrate well with equipment such as GPS either. Now, with an increased use of devices in the field and the use of drones to capture high-resolution imagery, such integration is increasingly important.

Today, you need only one piece of equipment to capture much of the data needed in the field: your smartphone. Data can be synced in real time with your GIS back at the office or in the cloud. These advancements have streamlined workflows, saved money and time, helped create a more geo-aware society, and increased the pace and effectiveness of land protection and management. Meanwhile, artificial intelligence (AI) enables quantities of data from satellites, sensors, photos, text, and more to be processed in a fraction of the time it would take one person or a team working in a desktop or small computing environment to accomplish.

Big data, advanced technology, infrastructure, and GIS are helping us solve problems that we couldn't before. So, what's next? How is the field changing, and what novel approaches should we track that could advance land protection faster and further?

As a result of the move from desktop to web- and cloud-based computing, we can now access GIS from anywhere and publish and share apps and maps to anyone with a web-enabled device—whether or not they are connected to the web or cellular service. This makes GIS more accessible to everyone from K–12 students to nonprofits. As Jack Dangermond, president and CEO of Esri, says, GIS is the "nervous system of the world" because it's so pervasive and so many people can now access it.

The skill sets of GIS professionals have broadened to include coding, statistics, data science, and more varied graphical or product-focused expertise. GIS professionals are now integrated into operations, advising on organizational issues and strategy, and producing and maintaining products, solutions, and data that is mission critical.

What follows are some examples of how organizations and companies are using technologies and GIS in creative and innovative ways to support and advance land protection.

Ocean science

Esri Oceans Hub brings the latest information on advances in ocean resources into one location. The hub includes links to new data, blogs, scientific research, ArcGIS StoryMaps, lessons, and more. Use it to stay up to date on the latest developments, such as the creation of advanced basemaps that give more detailed information on ocean depths and bathymetry (ocean floor topography).

Coral reefs have been bleaching and dying at an alarming rate because of climate change, human pressures, and other factors that we are only beginning to understand. To protect and restore these coral reef ecosystems, we need better data, maps, and science. Only a small percentage of reefs worldwide have been accurately mapped. But even for those, we've amassed little detailed information about reef biology, structure, or health. A partnership between The Nature Conservancy (TNC), Arizona State University's Center for Global Discovery and Conservation Science, and Planet is tackling this challenge for the entire Caribbean Basin (figure 10.1).

TNC has mapped shallow benthic habitats at different scales and used various imagery sources to answer management questions at regional, national, and local levels. By combining satellite and aerial imagery with field data from drones and underwater cameras, TNC has created the most detailed, high-resolution shallow benthic habitat maps to date. This data is informing marine spatial plans, coral restoration, ecosystem service models, climate adaptation projects, and much more. To learn more about TNC in the Caribbean's work, you can visit the Caribbean Science Atlas (figure 10.2).

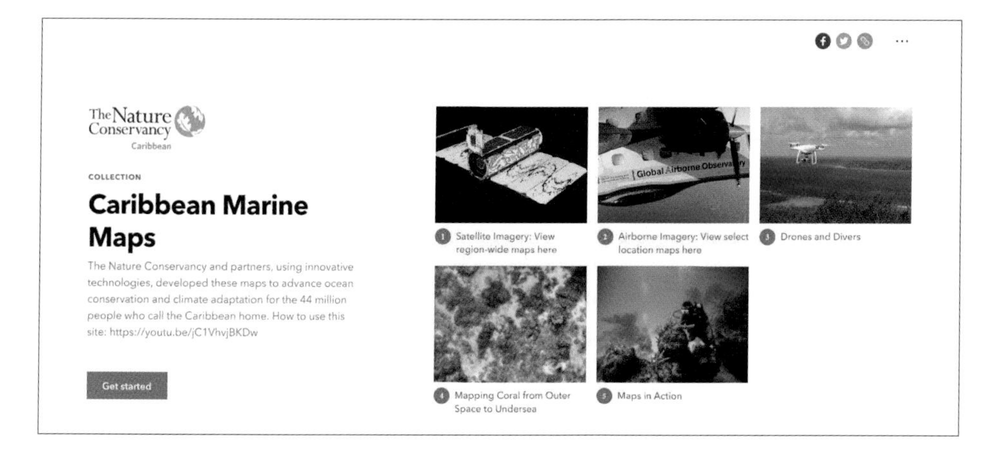

Figure 10.1. CaribbeanMarineMaps.tnc.org is a StoryMaps Collection that compiles all of TNC in the Caribbean's resources on mapping coral and benthic habitats at different scales. Users can interact with data layers directly, download GIS layers for analysis, and learn more about how and why these maps were created and how they are being put to work to inform conservation and restoration actions on the ground.

Valerie Pietsch McNulty, courtesy of The Nature Conservancy.

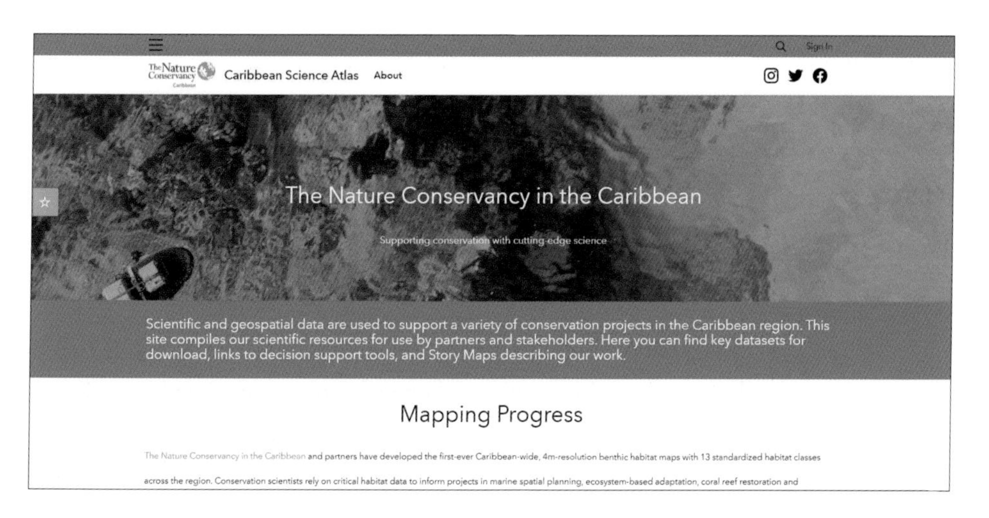

Figure 10.2. CaribbeanScienceAtlas.tnc.org is an ArcGIS Hub site that compiles scientific and geospatial products created by TNC in the Caribbean for use and review by partners and stakeholders in the region. Users can find datasets for download, decision support tools, and StoryMaps Collections with the latest news about ongoing conservation efforts.

Valerie Pietsch McNulty, courtesy of The Nature Conservancy.

Virtual and augmented reality

Virtual reality (VR) is the projection of an immersive computerized environment that users can interact with in space through a smartphone, desktop, or special VR headset. Augmented reality (AR), by contrast, doesn't transport you—it projects images on your screen that correspond to locations in space around you. (A popular use of AR is the game Pokémon Go.)

These technologies are on the rise among GIS professionals in conservation and parks. Both VR and AR can help visualize what a land-use change could mean for an area. For example, in figure 10.3, Defenders of Wildlife uses AR to show visuals of a landscape before and after the proposed removal of a dam. Defenders of Wildlife will use AR and VR to inform decision-making on a local level by helping people visualize a project landscape and the potential impacts of land-use changes

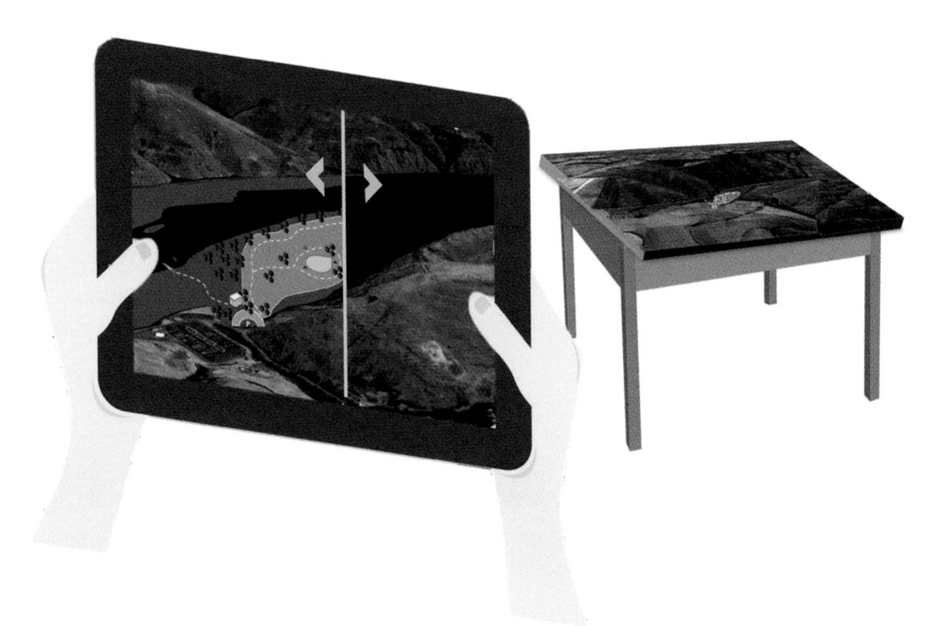

Figure 10.3. AR rendering of what the viewer will see on their phone/tablet close up and zoomed out. The 3D surface will be fixed to a table. The zoom in shows real river shoreline data before and after construction of a dam. Also included is a hypothetical park blueprint to show how people can start reenvisioning a future riverscape without dams.

Defenders of Wildlife Center for Conservation Innovation.

in that area. It is also using this technology to promote advocacy by helping people appreciate the importance of places they may never visit but can still act to protect.

In another example, the nonprofit Unique Places to Save created NatureXR to expand the use of immersive technologies to advance education, conservation, and restoration of the natural world. It is creating a version of scientifically certified "digital nature" by combining 3D scanning, remote sensing, and VR. Biological and content experts collaborate to ensure botanical accuracy. With the creation of trees and plants that mimic real structure, canopy, crown, and other features (figure 10.4), VR tree data can be used to help us understand how species function and how they respond to their environment. How does a certain tree species cast shade in an environment at a certain time of year with differing cloud cover? How will the trees grow and change in relation to increased temperatures now and into the future as the planet warms? Are cities investing in planting tree species that can hold up to drought and warming temperatures? The answers to such questions can help shape our restoration plans and designs.

Figure 10.4. NatureXR scientifically certified digital nature—an example of longleaf pine.

Unique Places to Save.

Integrating science-certified features with the power of GIS, AI, and other applications has much potential to expand our understanding of the intricacies of the natural world. It can also help us plan and design for a more sustainable future.

VR and AR have long been used in fields such as planning and defense but are becoming more affordable and accessible to the land protection community. Park departments are using VR to give people the opportunity to view different park designs or streetscapes. Land conservation specialists are using these tools to show how land-use changes would alter viewsheds. These tools are being integrated with GIS to understand how new buildings in cities would affect views from other properties or how street trees would shade a street at certain times throughout the day. VR is being used to immerse viewers underwater with whale sharks and in the jungle forest canopy with monkeys. You can virtually dive into the ocean and look up at the scale and extent of pollution from ocean plastics. VR and AR are already revolutionizing our ability to educate people with immersive experiences that can turn them into supporters and advocates for conservation and parks.

People mobility data for land use and space utilization

Advances in computing, AI, and machine learning have made it possible to access large amounts of data on the movement of people. This data is collected from mobile phone apps and anonymously identifies where people are traveling, how long they stay there, the time of day they travel, and so on. This data requires a lot of preparation, scrubbing, complex algorithms, and massive computing power to run the analytics. Many companies offer access to this data, but it's expensive, and most nonprofits don't have the money or expertise required to use it. But companies such as Resilient Solutions 21 (RS21) and Esri, and universities such as the University of Washington and University of Arizona, are working to make this people mobility data more accessible.

RS21 is using anonymized cell phone data to analyze use patterns of protected lands. RS21 worked with The Trust for Public Land to develop a project called VIBE. The project used anonymized cell phone data to understand where park users travel from, user counts by neighborhood, and demographic and socioeconomic characteristics of park users. The data also reveals where people aggregate in the park and how long they stay there. You can also look at the data by week, day, and time of day (figures 10.5 and 10.6).

There is so much promise for this data to uncover use and other trends for protected lands. Such data could unlock more funding for additional lands protection or management and demonstrate return on investment and ecosystem service benefits. People movement data verified the exponential uptick in park and open space use during the COVID-19 pandemic. What else can anonymized cell phone data be used for in our field? The opportunities are wide open to link data such as this with data science, research, and analytics to help solve real-world land protection and management issues.

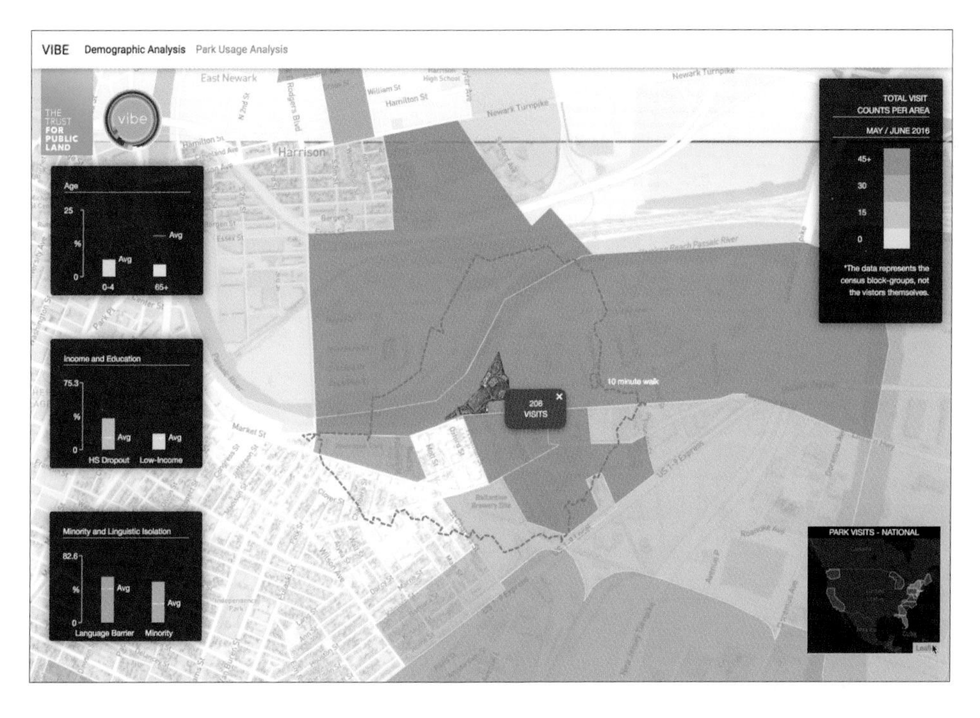

Figure 10.5. VIBE mobility data app shows park visits by surrounding census block groups with socioeconomic characteristics.

Resilient Solutions 21.

Figure 10.6. VIBE mobility data app shows the location of devices that entered the park geofence during a two-month period. The app enables the visualization of park visitor movements by day and time series.

Resilient Solutions 21.

Protected area management solutions

We have long needed to democratize advanced, enterprise-class systems to support the scaling of land protection. Organizations keep trying to scale with traditional technology approaches and struggle to meet the challenges at hand. Land protection is a complex activity that requires the integration of monitoring and managing diverse topics. Few organizations have the technical staff to support the integration effort required to achieve the complete system configurations required for holistic land protection. However, a new pattern is emerging to lower the burden of system design and implementation to support protected area management and community-led conservation programs (figure 10.7).

Figure 10.7. African People & Wildlife team members use GIS to capture the point location of a paw print of an animal.

African People & Wildlife/Felipe Rodriguez.

Esri and the National Geographic Society (NGS) have partnered to develop simplified "solutions" that can be easily deployed to support protected area operations, real-time management, and analysis. These solutions come preconfigured with a suite of apps that work seamlessly together to support primary protected area management activities. Lowering the barriers to implementation, these ready-to-go solution toolkits include customer support to ensure successful use.

ArcGIS® Solutions for Protected Area Management, powered by ArcGIS Online, includes workflows and apps to support protection, management, and monitoring. The solutions currently include protection operations, wildlife management, and conservation outreach. A set of apps and processes helps track and stop poaching with the use of wildlife movement data. The apps enable monitoring and reporting on the status of wildlife populations. Apps focused on community engagement collect data on human/wildlife interactions and illegal activities within the protected areas. These apps can be used alone or bundled and are integrated into a command-center dashboard (figure 10.8).

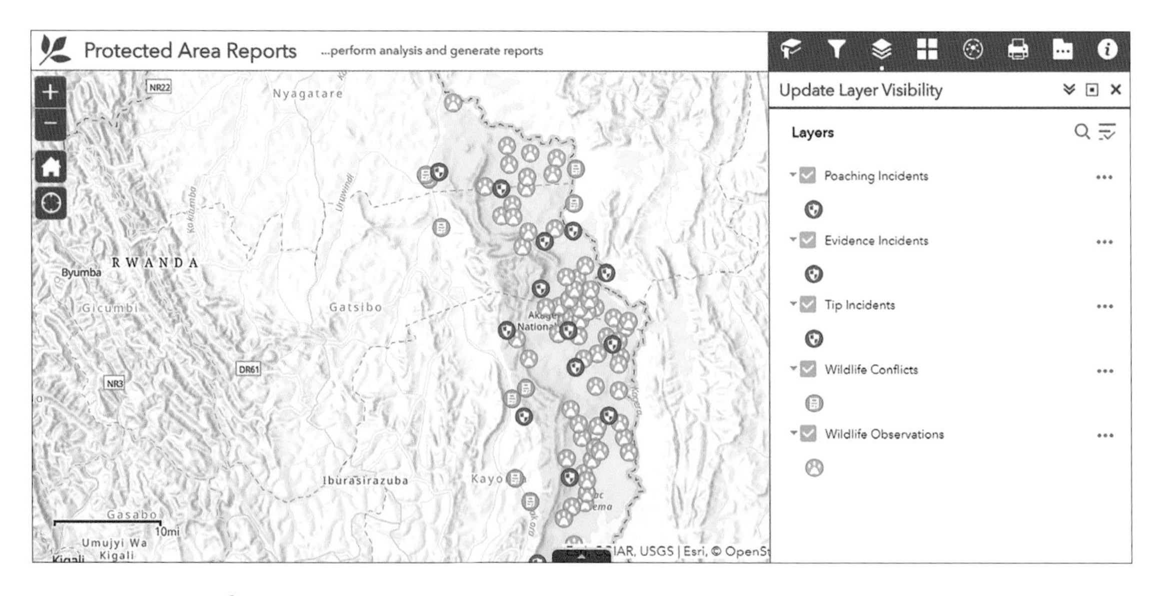

Figure 10.8. ArcGIS Solutions for Protected Area Management Reports.

Having all this data linked together in the dashboard enables rapid deployment of security staff to stop poaching or illegal activity in the park. It provides a platform to bring data out of silos and into one real-time display of what is happening across hundreds of thousands of acres of protected lands with varied terrain. This platform supports strategic land protection decision-making from one centralized solution that all park staff can access and use—a novel approach to expanding the availability and accessibility of a complex solution made simple for real-time operations. View the *Managing Protected Areas* and *Getting to Know the Conservation Solution* StoryMaps stories to learn about how partners such as Africa Parks, African People & Wildlife, Peace Parks Foundation, and the Jane Goodall Institute (JGI) are using these solutions with great success (figure 10.9).

JGI and Esri are partnering to develop another solution to support community-led conservation. This effort aims to unlock the community-led conservation engagement pattern Jane Goodall crafted while working with communities surrounding Gombe National Park in Tanzania, known as Tacare. Generally, these user stories are focused on lowering the burden to collect data from communities and ensuring that collected data is easily shared. The configuration of apps supports the following community planning and engagement workflows.

Figure 10.9. Rangers using the ArcGIS Solutions for Protected Area Management app.

Apps supporting community planning and engagement workflows

- Community Context—summarizing information within a given distance of a community of interest, enabling users to access readily available ArcGIS Living Atlas of the World content and other conservation entity–specific data layers (e.g., collected demographic data)
- Facilitated Sketching—collecting location-based information of significance shared through paired learning experiences within communities
- Stakeholder Mapping—documenting stakeholders within a community, measuring their level of support and participation to understand power dynamics within the community

These initiatives represent a new pattern to advance and broaden the use of GIS for conservation impact by using turnkey "conservation solutions" to lower the barriers of entry. Go to the ArcGIS Solutions for Conservation home page for updates on these offerings. Figure 10.10 shows a white rhinoceros, a near-threatened species.

A groundbreaking and promising example of pairing machine power through AI and people power through decision-making and action is the use of motion-sensing cameras and image detection. Many of the world's protected areas span tens of thousands of square miles but are patrolled by only a handful of people. RESOLVE, an environmental nonprofit, teamed up with Intel, NGS, Arrow Electronics, the Leonardo DiCaprio Foundation, and other partners to develop TrailGuard AI cameras. These cameras are equipped with Intel processing chips and tied to cloud apps and services that are trained to process hundreds of thousands of images to detect potential poachers and alert rangers immediately. Using these devices has led to the arrests of 30 poachers and the seizure of more than 1,300 pounds of bushmeat

Figure 10.10. A white rhino in Kenya. The white rhinoceros is on the International Union for Conservation of Nature Red List of Threatened Species as near threatened with population declining.

RESOLVE.

during the first field test at the Grumeti Game Reserve in Tanzania. RESOLVE is working to scale the technology and make it affordable and accessible to support wildlife and land protection efforts worldwide (figures 10.11 and 10.12).

Figure 10.11. Humans detected by the TrailGuard AI anti-poaching camera system at a game reserve in East Africa. Faces hidden for confidentiality.

RESOLVE.

Figure 10.12. Elephants walking in Kenya.

RESOLVE.

Equity and environmental justice

Data science is increasingly being used to tackle social issues and improve the quality of life for communities around the world. Many organizations and companies understand the value and benefit of applying data and insights to social issues to help solve these issues faster. Data science is used in the private sector to understand issues such as customer behavior, marketing strategies, and health-care interventions that drive innovation and new business practices, but it hasn't been accessible to the social and environmental sectors—until now. Many organizations are innovating in this space. GiveDirectly is a nonprofit that makes direct money transfers to the poorest households in rural Kenya and Uganda. They needed a better process to identify where the poorest households are located. Thatched versus metal roofs are a proxy for poverty. They partnered with DataKind to use machine learning and image processing to identify villages and households with thatched roofs to help make their giving more efficient and effective. Though there were some challenges when data was ground-truthed, this type of innovation is a great example of how more organizations are turning to technology to support social change.

Data.org brings people and organizations together to effect positive social change and build the field of data science for social impact. The platform links nonprofits with public, academic, and private partners to solve some of the world's most challenging social problems. For example, finalists in the 2021 100&Change grant are using data and technology to address issues such as homelessness, stopping the spread of mosquito-borne disease, and eliminating "news deserts." In the conservation and park field, there is much room for growth and expansion in using data science for equity and environmental justice. Our failures to meet land protection goals will disproportionately affect low-income communities of color, indigenous communities, and people in the developing world. But we can harness the power of data science and advancing technologies to analyze and address these issues before they happen—to solve them faster and smarter.

These are examples of groundbreaking uses of technology and GIS that make it possible to share data, tools, apps, and processes so every land protection organization or agency can access them—and not a moment too soon. Since I started writing this book in late 2019, data and science tell us that we are not on the right trajectory to save our planet, and we must course-correct now. Polar bears and certain species of whales are on a course to be extinct by 2100, along with two-thirds of the bird species in America. COVID-19, an infectious zoonotic disease, has swept across the globe. As humans keep pushing into the last wild places on earth, scientists and public health experts expect more of these infectious diseases to emerge.

All these disasters can be slowed, or even reversed. GIS and emerging technologies are key to meeting these challenges, and it is up to us to scale our collective land protection efforts now. We can course-correct.

I once asked a friend how she keeps hope alive when there are so many heartbreaking, devastating, infuriating, and seemingly impossible odds stacked against us and the natural world we love. She said she didn't think it was hope that would solve our problems—it's our determination. We must set our sights on the actions that each of us can contribute to preserving the places we love and act with determination to innovate our way through the chaos to save our world. Figure 10.13 shows a "rainbow of hope" over the Galisteo Basin Preserve.

I maintain hope and am determined to use technology and GIS to turn the tide. Writing this book has given me the opportunity and gift to connect with your work, your visions, and your creative uses of data, science, and technology for land protection. I am deeply inspired by all your efforts, and I can tell you, you are not

Figure 10.13. A rainbow of hope over the Galisteo Basin Preserve, New Mexico.

alone. We are a passionate and incredibly smart community working toward the same goal of protecting and sustaining our world. We are collectively contributing to the visions such as the Half-Earth Project, the United Nations' Sustainable Development Goals, and the Thirty by Thirty Resolution to Save Nature. But we must do more, faster and more efficiently, with the technology available to us. We must reach out to our neighbors and our colleagues in other organizations to minimize the crossover and duplication and to amplify our efforts. We must not be afraid to stretch our minds, our partnerships, and our technologies to protect and restore the places we love, both close to home and far away.

If you are new to GIS, find a peer, volunteer, or consultant who can help you get started. If you use GIS and after reading this book have more creative ideas for how to achieve your land protection goals faster, put those ideas into action right away. Find the right partners and build new relationships. Make it happen. If you are an expert user, identify organizations in your sphere of work, geography, or influence that could be using GIS but aren't. Reach out to them with guidance and support. Share, share, share. Share data, share tools, share expertise. Data and knowledge silos only slow down our ability to save our world. Democratizing technology, data, and GIS is a critical part of the solution to the challenges we face now. We have no time to waste. It will take all of us to work together for a future where all places and all species, including us, are protected, connected, and thriving.

Index

B

C

D

E

F

G

H

I

M

N

O

P

Q

R

S

T

U

V

W

Y